CHANNELING CULTURES

Television Studies from India

edited by
Biswarup Sen
Abhijit Roy

OXFORD
UNIVERSITY PRESS

Oxford University Press is a department of the University of Oxford.
It furthers the University's objective of excellence in research, scholarship,
and education by publishing worldwide. Oxford is a registered trademark of
Oxford University Press in the UK and in certain other countries

Published in India by
Oxford University Press
YMCA Library Building, 1 Jai Singh Road, New Delhi 110 001, India

ISBN-13: 978-0-19-809205-6
ISBN-10: 0-19-809205-9

Typeset in Trump Mediaeval LT Std 9.5/13
by Alphæta Solutions, Puducherry, India 605 009
Printed in India at G.H. Prints Pvt Ltd, New Delhi 110 020

Contents

Foreword

What started life some years ago as an international seminar at the Indian Institute of Advanced Study, Shimla, with the title '50 years of Indian Television: Contemporary Issues', is now being published as a book under the more attractive title 'Channeling Cultures: Television Studies from India'. This small change in title I would like to over read here. The packaging must change when the audience changes. There are good reasons for it. The earlier title, staid and descriptive as it was, was directed at institute administrators who were paying for the event but it would not travel too far for a paying public. Something more suggestive and attractive was required and thus the later title. Both titles have the same strategic element to them, of getting the attention of the intended audience. Both will have served their purpose. Further, in addition to such strategic positioning, there is another crucial element that we must factor in when publishing the book: time. Since it is coming out some years after the seminar, when the papers have gone through a rigourous review process and the authors' perspectives have perhaps changed, as a result of the new debates that have entered the discipline, the book that is being published has become very contemporary. You can sense it in the change in title. By introducing the word 'channeling', with its double meaning, one of which signifies the impatient viewer spoiled for choice as he/she surfs channels to find something interesting, and the other the astute producer concerned with directing eyeballs to his/her channel, the title signals what is to follow in the book. Television has today to compete for attention with the possibilities that have emerged as a result of the new digital technologies which have given us websites such as YouTube and Facebook. Television is under challenge. This book is about television's story in India.

If we look at the years between the seminar and the book, from this perspective of competition for viewers, it seems as if we are looking at an age. We could even regard it as being equivalent to the previous fifty years of television measured in terms of the revolution of expectation and behaviour that the digital technologies have produced. The public discourse too has changed. In the first three decades television's objectives were different. Television was engaged in a different set of policy trade-offs and it is necessary for us today, when we review the fifty years, to be reminded of them. In an earlier period television was an instrument of nation building. This was consistent with the post-colonial discourse of the time, about the building of the new nation's capacity for independence summed up in that evocative phrase 'we must be self reliant'. Television was to be judged in terms of how far it served this goal, how far it made our thoughts and perceptions self–reliant. The discourse has changed and today we talk about television as part of the public right to leisure, as national gossip, and as a national timepass. The story of fifty years of television must therefore be told in terms of this changing discourse, the elements of which are the new technologies of broadcast, its function as an instrument of state, its role as a medium of instruction, its service as a channel of entertainment, and its emergence as unaccountable power masquerading as the fourth estate. Television in India has many personalities. Over the years it has also created new publics which have both private and public interests. Many of these aspects have been addressed in the interesting chapters of the book. Being one of the first collaborative books on television in India it is a valuable resource not just for recording the story of television evolution but also for formulating the questions that media studies must ask. In these fifty years television has travelled across many time zones and this journey mirrors the travel of the nation.

Permit me some reminiscences here. I belong to a generation that can remember an India when television first appeared. I was born into a middle–class family where buying the first black and white television set was a momentous event. We had to bear in mind that not only were we buying a new technology but we were also buying an elegant piece of furniture for the drawing room. German manufactured televisions, Grundig was the brand, came in a wooden frame, varnished and with a good lacquer finish. The decision to buy was preceded by weeks of considerable planning and followed by days of considerable celebration. In the drawing room the television stood and had for us children a hallowed presence. It gave us bragging rights in school. Programmes were educational and educational programmes entertained. It was less of the idiot box than it is today when clones of American popular shows are aired. The films shown were what the Films Division made available. Leisure time had to be re-arranged. The all-important order which we as children waited for, 'go out and play', and which defined childhood, had its stature diminished by the subsequent challenge 'but I want to watch TV'. The challenge became the new monarch. What was true for the middle class then, I am sure, still remains true for the rural poor today even though the content of what is broadcast is now of a different order. We watched cricket then. We also watch BPL (not below the poverty line but Barclay's Premier League) football today.

Television in those days was about nation building. The farmer and his needs were paramount, with weather forecasts, cropping patterns, new technologies of tilling the soil, water management, pest control, and so on, being broadcast. The farmer was an important constituency and an important consideration of state. National pride was an important goal and so, the achievement of our scientists with satellite launches, atomic implosions for peaceful purposes, and the green revolution were all given key broadcast slots. The inauguration of the temples of modern India received prime time attention. Television programmes, such as Satellite Instruction Television Experiment (SITE) and APPLE, that had an educational goal, and an all-India footprint because they used satellite transponders, were launched with much fanfare. Having our own space programme, one goal of which was to educate the nation, was considered necessary and so when Vikram Sarabhai the champion of the space programme was asked whether we as a nation could afford such expense he countered by saying 'Can we afford not to?'

Nation building was a slogan and a credo. While it may be infra dig (not a term in much use today) as an idea today, it was necessary in the 1960s and 1970s as the country was emerging from the anxiety of a peaceful leadership transition, a period of severe food scarcity, two wars, and was moving towards becoming a more confident nation. Television was growing as a tool of nation building. Having built the nation the discourse has now changed to seeing this period as one where television was a tool of governance. In those early decades there were only government broadcasts. The Government was the sole broadcaster and the government decided what was good for the nation. Doordarshan, a wonderfully evocative name for a television channel, came into existence. It educated, entertained, enabled, and empowered. It also employed, and employed many, as it grew into a powerful public institution. It was unable to shake off the work culture of all public institutions. In fact it would be interesting to study the sociology of this public institution because it would illustrate the tensions that must emerge when a revolutionary technology gets encased in a public institution that has public goals and an operating culture no different from other public institutions. Doordarshan, because it has a financial safety net for itself and secure employment for its staff, is in a unique situation that gives it autonomy from the vagaries of market forces. Given these aspects, an ethnography of its 'darshan' is just waiting to be written.

Reflecting this changing public discourse, on the function of television in a developing society such as India, I recall one episode which to me now appears definitive. Since remembering things past constitutes an archive of memory, and since such an archive is important for us to know where we have come from and where we wish to go, I want to use this foreword to add my own memories to the fascinating accounts in the chapters of this book. Television in India needs such an archive and this book should be seen as initiating its construction. It is a record of television's memories. In the 1970s, the idea of development dominated our thinking. We talked about development and underdevelopment, about the New International Information Order that would emerge along with the New International Economic Order. Central to such discussions was the issue of terms of trade, between countries, between regions within a country, and between sectors. Resource allocation was at the centre of our arguments and in that euphoria of nation building we were still prepared to give the farmer priority in allocation of public resources over the demands of the middle classes. The self did not enjoy the pre-eminence that she enjoys today. Being selfless, for the larger public good, was the message of the many moral science classes that we were all required to attend. This is what nation building required.

In Delhi in the 1970s, some young faculty members at Indian Institute of Technology Delhi and Jawaharlal Nehru University started a movement called SPIC MACAY (Society for the Promotion of Indian Classical Music and Culture amongst Youth). One of the activities of the movement was to organize a lecture demonstration which involved inviting a renowned artiste to speak to students and to demonstrate, through a performance, what she was describing. On this particular evening Yamini Krishnamurthy the renowned Bharatnatyam and Kuchipudi exponent was the artiste in question. The chief guest was the Cabinet Minister of Information and Broadcasting, Shri Vasant Sathe. There was a robust public discussion going on in the papers on whether we should shift from black and white to colour television, on whether we had the resources to do so. The main concern was about costs and about the diversion of scarce resources from areas such as irrigation to setting up the technology for colour television. I remember being on the side of black and white as against colour, adopting the high moral ground of sacrifice for the public good. When Shri Sathe in his opening remarks said 'can you imagine seeing Yaminiji in black and white?' we knew the new order had arrived. We couldn't imagine her in black and white, not her resplendent silks, not her make-up, and not even the flowers in the hair. On that day 'culture' which had been waiting in the wings of national policy trumped 'development'. Colour television heralded a change in the public discourse. It was a shift in the locus of public power. The Minister's rhetorical statement became a policy pronouncement on the country's decision to spend money on building the infrastructure for colour television. Culture has now become popular culture as the enlightened state has yielded space to the pragmatic market. The middle classes announced their arrival, both as social fact and as social aspiration.

In looking at fifty years of television in India we are also looking at a changing public discourse. The medium and the messenger, and the metaphor of shooting, have all come together. Shooting the messenger, who is shooting her mouth off thereby causing people to shoot themselves in the foot, is what television has become in its breaking news broadcasts. Television has become the new sacred cow of Indian democracy. Because it presents itself as the fourth estate, its excesses are condoned and its power is unchallenged. But the questions do not go away. What is its business model? What is its ownership structure? Should cross media ownership be allowed? Should corporate houses be permitted to own television channels if they have other business interests? Should they be self-regulated or should there be external regulation by some independent broadcast authority constituted by law? Should this independent authority have quasi-judicial powers that citizens can appeal to, if they have grievances against the television channels? A democratic society requires that these questions be asked. Democratizing Indian democracy requires us to look critically at one of the most powerful institutions to have emerged in independent India. Doing so is not shooting the messenger. It is in fact saving the messenger from an incipient tyranny. The public discourse in these fifty years has changed from promoting development, to providing cultural entertainment, to democratizing democracy. Television studies must accompany this changing discourse. The book is an acknowledgment of this responsibility. I am delighted that the Indian Institute of Advanced Study and Oxford University Press are co-publishing the book. I am grateful to the editors for making it happen, for carrying the burden of taking a good idea through the many stages from conception to creation. It is a valuable addition to not just media studies, or India studies, but also to comparative studies in democracy. This is most welcome.

PETER RONALD DESOUZA
Director
Indian Institute of Advanced Study, Shimla
31 July 2013

Acknowledgements

This book grew out of a conference on '50 Years of Indian Television', convened by the editors of this volume and held at the Indian Institute of Advanced Study (IIAS), Shimla, in July, 2009. We want to express our deepest gratitude to Professor Peter Ronald deSouza, the then Director of IIAS, for agreeing that a conference highlighting developments in Indian television studies was long overdue. A gracious organizer and host for the event, Professor deSouza took the initiative with Oxford University Press (OUP) in making this publication possible. Given his involvement in the project and his critical interest in the media, no one could be more fitting for writing the Foreword to this volume. Special thanks are due to the administration and staff of IIAS for the hard work they put into organizing the conference and for making all the participants feel so much at home in Shimla. We would also like to acknowledge our debt to Bhaskar Ghose, Paranjoy Guha Thakurta, Ujjwal K. Chowdhury, and Bhaskar Das for participating in the conference and providing valuable insights drawn from their distinguished professional experience in journalism, media industry, and the government. We thank OUP's entire editorial and production team for providing valuable inputs and assistance throughout the publication process, and for being so easy to work with. We remember with gratitude the efforts put in by Urmila Dasgupta and

Neha Kohli of Oxford University Press at an earlier stage of the publication process. We are deeply indebted to Professor Arvind Rajagopal for agreeing to write the Afterword for this volume. Finally, we are grateful to all the contributors for not only making the project what it is, but also for their patience and forbearance while waiting for the volume to come to fruition.

Abbreviations

AIADMK	All India Anna Dravida Munnetra Kazhagam
AIDS	acquired immunodeficiency syndrome
AIR	All India Radio
BJP	Bharatiya Janata Party
BPO	business process outsourcing
C&S	cable and satellite
CCTV	closed-circuit television
CEO	chief executive officer
CFAR	Centre for Advocacy and Research
CPI(M)	Communist Party of India(Marxist)
DART	Doordarshan Audience Ratings
DD	Doordarshan
DDS	Deccan Development Society
DG	Director General
DMK	Dravida Munnetra Kazhagam
DTH	direct-to-home
DVR	digital video recorder
EPG	Electronic Programme Guide
FDI	foreign direct investment

FICCI	Federation of Indian Chambers of Commerce and Industry
IMF	International Monetary Fund
IPL	Indian Premier League
ISRO	Indian Space Research Organization
IT	information technology
JD(S)	Janata Dal (Secular)
KCP	Kheda Communication Project
KMVS	Kutch Mahila Vikas Sangathan
KSEB	Kerala State Electricity Board
LGBT	lesbian, gay, bisexual, and transgender
LTTE	Liberation Tigers of Tamil Eelam
MP	Member of Parliament
MPA	Media Partners Asia
NASA	National Aeronautics and Space Administration
NBER	National Bureau of Economic Research
NDTV	New Delhi Television
NGO	non-governmental organization
NRI	non-resident Indian
PMK	Paattali Makkal Katchi
PMO	Prime Minister's Office
PRB	Population Reference Bureau
PwC	PricewaterhouseCoopers
RSS	Rashtriya Swayamsevak Sangh
SITE	Satellite Instructional Television Experiment
S–M–C–R	source–message–channel–receiver
SMS	short messaging services
SOAS	School of Oriental and African Studies
TAFE	Tractors and Farm Equipment Limited
TDP	Telugu Desam Party
TRP	Television Rating Point
TVR	Television Viewer Rating
UK	United Kingdom

ULFA	United Liberation Front of Assam
UNCTAD	United Nations Conference on Trade and Development
UNESCO	United Nations Educational, Scientific and Cultural Organization
US	United States
USP	unique selling point
VSNL	Videsh Sanchar Nigam Limited
WAN	World Association of Newspapers

Introduction

BISWARUP SEN *and* ABHIJIT ROY

One of the most notable political events in recent years—the confrontation between the Indian state and an agitated 'civil society' over the question of an anti-corruption bill—was memorable for playing itself out twice: once, in real life; and a second time, on the television screen. The instant replay of the 2011 Jan Lokpal controversy also provided an instant snapshot of Indian television today: footage of the proceedings at Ramlila Maidan underscored television's status as a live document of contemporary history and as the arbiter of truth; interviews soliciting the views of the man in the street as well as of luminaries from high places demonstrated its extraordinary capacity to represent the complex multiplicity of the nation's citizenry; the stance adopted by anchors and commentators signalled the moral–political agency of the emergent middle classes; the overload of visual information and the frenetic mode of speech echoed the beat of contemporary market-driven broadcasting; and the constant feedback loop between (Short Message Service), social media, and the television demonstrated television's convergence with new media technologies. In short, television provided the frame as well as the ingredients that made this historic face-off between the government and civil society possible, thus establishing its crucial role as a facilitator of political and ethical debates in contemporary society. As one of the contributors to this book points out, 'it is impossible to imagine, or explain, modern India without reference to television'.[1]

Television as a medium has a mammoth presence in the current Indian media environment, occupying 45.19 per cent of the media and entertainment industry in India in the year 2011. India is today the world's third-largest television market, with almost 146 million TV households; the number of TV channels in India went up to 623 in 2011, and is growing every year. According to current projections, the sector is poised for a 17 per cent compound annual growth rate till 2016, to touch annual revenues of approximately Rs 735 billion.[2] This phenomenal growth, it needs to be pointed out, has occurred in the context of incredible lags in other areas of social development. Thus, while television penetration in India was approximately 60 per cent of total households in 2011[3] (and a similar number have access to telephones), less than 47 per cent of households have access to an indoor toilet.[4] What is important to note here is not merely the familiar fact of 'uneven development' or even the increasing currency of information and entertainment as 'basic' needs of life. What we witness here is a version of modernity in which the spread of communication technologies yields unpredictable results that subvert simple causal accounts of the relationship between communication and growth. While new communication technologies can open up discursive spaces that promote equity and egalitarianism, they are equally capable of injecting a new lease of life into old habits and practices, thus rendering a greater visibility of existing social divisions.[5] It is from such an understanding of television's nuanced location in postcolonial society that a phenomenon like Anna Hazare's anti-corruption movement, and the participatory culture triggered by it, becomes crucial for television scholars.

This book grew out of a conference on '50 Years of Indian Television' held at the Indian Institute of Advanced Study (IIAS), Shimla, in July 2009. Participants in the conference—many of whom are contributors to this volume—presented papers that examined contemporary trends in Indian television, while focusing on the historical contexts that could help us understand these developments. At a moment when the Indian state is seeking to alter the television landscape with a number of initiatives—for example, compulsory digitization, directives limiting the duration of advertisement per hour, and the proposed code for content regulation—and when 'media ethics' has become a serious topic of debate in the public domain, it is an opportune time to present a book that will hopefully stimulate a rethinking of the role of television in the cultural politics in India. Indian television has, in fact, gone through a series of rapid changes since its inception in 1959 as an educational tool designed for social development: the disenchantment with exclusively development-oriented programming after the Satellite Instructional Television Experiment in 1975–6; widespread commercialization starting with the 1982 Asian Games; early concession to entertainment with the introduction of films and later, film-based programmes like *Chitrahaar*; the era of the socially conscious soap operas (*Hum Log, Buniyaad, Rajani*); the extraordinary impact of *Ramayan* and *Mahabharat*; the birth of popular interactive television shows like *Surabhi*; the coverage of the Iraq war by CNN in 1991; the coming of private operators like STAR and Zee and the quick development of a multichannel environment; the redefinition of the notion of 'public' by the historic Supreme Court judgement in 1995; the formation of the supposedly autonomous public media authority called Prasar Bharati; the introduction of new modes of television advertising; the birth of 24-hour news channels pioneered by NDTV; as well as the huge presence of Bollywood and live cricket. These developments have matured over the years and have, in turn, produced new trajectories in Indian television: reality TV; sensational news coupled with allegations of 'trial by media'; widespread participatory culture through such mechanisms as call-ins, SMS, and web-based feedback; the increasing importance of small towns and the regional media; complex patterns of industrial cross-holdings; the huge popularity of 'spiritual' television and reality talk shows engaged in truth finding and activism; panel discussions on current affairs; a wide range of sports from football to Formula 1; fashion and lifestyle; factual popular and stand-up comedy; and the rise of infomercials promoting a wide assortment of goods and services for sales. This vast array of content has begun to acquire a global audience; for example, the dubbed version of *Kyunki Saas Bhi Kabhi Bahu Thi*, a popular Indian television serial, has become a huge hit on Tolo TV, the premier television channel in Afghanistan; Zee TV has seen significant growth in Europe; and Indian channels are acquiring loyal followings in many parts of Asia and Africa (see Dudrah 2002).

The book, by looking at a wide cross-section of current trends—the impact of satellite television on television news, the increasing coverage of Bollywood and sports on news shows, the rise of reality TV, gender politics in soap operas, the peculiarities of news production in regional television, and the coverage of 'terrorism' on national news channels—hopes to shed light on a number of questions that are of central concern to media scholars in the subcontinent. How can televisual form be approached from the contexts of postcoloniality and globalization? How do we characterize the post-liberalization era of Indian television? Are the lines between news and entertainment getting increasingly blurred under the pressures of market forces? Is regional television a smaller version of its national counterpart or does it have distinct features of its own? How do genres like soap operas and reality TV provide a space for the articulation of the global and the local?

Television is the medium that most adequately represents the various territorial levels that make up the totality of modern Indian society and thus constitutes a privileged object of inquiry for media scholars. Staging a complex interplay of various geopolitical levels—local, regional, national, transnational (the participation of Pakistani media personalities on reality TV, for example), and the global (Discovery shows dubbed in the vernacular)—it enables an analysis of the complex ways these societal layers are constituted and related. The interconnected histories and temporalities that bind these levels become a major point of consideration for the chapters in this volume, each of which tackles a particular aspect of television today in order to broaden our understanding of contemporary Indian modernity.

As the editors of *De-Westernizing Media Studies* noted, their project was 'part of a growing reaction against the self-absorption and the parochialism of much Western media theory. It has become routine for universalistic observations about the media to be advanced in English-language books on the basis of evidence derived from a tiny handful of countries' (Curran and Park 2000: 3). This volume is in agreement with such a sentiment and intends to enlarge the scope of the field of television studies by interrogating the implicit universals that inform much Western media theory. Such a mission is not easy to accomplish in spite of the fact that television studies has expanded its focus quite substantially in recent years. In the words of a contemporary scholar, 'the discipline of global television studies is at once necessary and impossible: necessary because only global studies of television will reveal the multiplicity of television cultures that is critical for fighting the universalizing tendencies in Western discourse; impossible because any global study of television in the current geopolitics of international communication necessarily means an unequal discourse' (Kumar 2003: 137). We hope though, that by providing a rich analysis of television in a non-Western context, this volume (along with others in the same vein) will make an intervention towards the creation of indigenous perspectives in the discipline.

Television studies as a field was initially dominated by the tradition of mass communication research that was developed in the United States (US) in the 1940s and the 1950s. This positivist approach subscribed to the protocols of quantitative social science, and was characterized by methods such as content analysis, surveying, and statistical testing, and by the desire to construct predictive models of media use and effects. A substantial amount of work done in India is located with this tradition of mass communication research. Though we acknowledge the valuable data yielded by this mode of inquiry, we need to be aware of its limitations. As Arvind Rajagopal (2001: 337) has perceptively pointed out:

Indian media studies has tended to be dominated by an often overly empirical approach, where investigation into audience responses is thought to be capable of arriving at truths about the character of public opinion, of popular beliefs underlying national ideologies, and/or of subaltern sentiments whose exclusion can be compensated by a method of enumeration. Such an approach ignores the process of representation as well as the phenomenological specificity inherent to the work of a given media technology.

In other words, positivist inquiry can give us a precise description of the status quo, but can neither provide an account of how it came to be nor offer us alternative visions of media and society. One can extend Rajagopal's critique by observing that such research typically reinforces an epistemological distinction—between the West as the producer of 'theory' and the global South as a site of empirical demonstration—that needs to be interrogated and dismantled.[6]

The challenge to the positivist paradigm emerged from two different quarters in the 1960s and the 1970s. The works of Herbert Schiller (1971) and Dallas Smythe (1981) (amongst others) led to the founding of a 'political economy of communications' that drew largely from Marxism to focus on three crucial issues: (a) the processes by which audience power was created as a commodity that helped generate surplus value; (b) the structures of media ownership and distribution that guaranteed the stability and profitability of a capitalist system of entertainment; and (c) the manner in which Western (and more specifically American) media was exported around the globe as an instance of cultural imperialism. This perspective has since been further refined in the work of several Western scholars to address issues of ownership, control, and power in the era of global media.[7] While the economic underpinnings of Indian television (and more generally Indian media) has received some attention,[8] given the rapid capitalization of India's media industries and the wholesale transformations of business practices, the political economy of Indian television needs to be studied in greater detail.

This book can be situated within the various strands of political economy and cultural studies. All the contributors, no doubt, acknowledge the economic logic that governs capitalist media, and at the same time explore the ways in which television contributes to the politics of culture. The theoretical framework that informs the book needs some elaboration here. First, the domain of the cultural in contemporary capitalist societies is as important as that of the economic in determining the nature of the social formation. That is so because culture is not a direct reflection of the economic base as traditional Marxism had posited, but rather a relatively autonomous level that is (though not only) structured by its own causal mechanisms. Second, it is in the realm of popular culture—a space that is expansive and corresponds to a 'way of life'—that the efficacy of such an approach is to be located. As a corollary, an understanding of mass media is crucial to our analysis of the social whole. Third, culture is a site of ceaseless political struggle for hegemony carried on by semiotic means, that is, through a series of significations and counter-significations.

Beginning in the 1990s, cultural studies made a large impact on Indian media studies. The enormous vitality and reach of Indian popular culture meant any programme of inquiry which prioritized the study of media would find a very receptive audience amongst Indian scholars. Thus, much of the work on Bollywood published in recent years draws on its concepts and methods to explain the role of cinema in modern India. Moreover, the extraordinary diversity of India's political, social, and cultural landscape meant that a perspective which valourized difference as an irreducible ontological category would find fertile territory for academic exploration in the multitude of voices speaking through newspapers, film, radio, and television. Finally, the critique of traditional left thinking—mounted by, amongst others, the subaltern studies movement—meant that an epistemological space was cleared for a new kind of critical discourse that amounted to, in Stuart Hall's (1986) words, 'Marxism without guarantees'.

In the early years of its history, writing on Indian television consisted mainly of bureaucratic reviews of public television, statistical data, policy analysis, and recommendations (see, for instance, Chatterjee 1991). The past two decades, however, have produced a significant number of publications that approach television from a variety of perspectives. One of the very first examples of this new trend, published in 1990, took a critical look at the way women had been represented on Doordarshan (Krishnan and Dighe 1990). Ashish Rajadhyaksha's essay, 'Beaming Messages to the Nation', published in the same year, took a polemical view of the welfare state's notion of audiovisual literacy and speculated on the new possibilities available to art in the era of television and video (see Rajadhyaksha 1990: 41–2). Television's remarkable entry into popular culture from the mid-1980s also prompted journalists to take interest in the role of television in reshaping culture, and their writings provide a rich documentation of the myriad ways television had begun to impact everyday life in India (see, in particular, Kohli-Khandekar 2010; Ninan 1995). The first book-length study from the perspective of cultural theory—Ananda Mitra's *Television and Popular Culture in India: A Study of the Mahabharata* (1993)—was followed by a series of important publications on Doordarshan's crucial role in Indian popular culture. Nilanjana Gupta's *Switching Channels* (1998) probed deep into the troubled relationship between the missions of public television and the imperatives of globalization. Purnima Mankekar's *Screening Culture, Viewing Politics* (1999) provided a dense ethnography of perceptions of urban Indian women about television serials and their familial–moral universe, thus locating gender issues at the centre of narratives of the Indian nation. Arvind Rajagopal's (2001) highly influential work, *Politics after Television*, was exemplary in the way it balanced field surveys with media and cultural theory in an effort to show how television had changed the context of politics in India. Two significant journal issues devoted to Indian television—one, *Journal of Arts and Ideas* (1999); and the other, *Journal of the Moving Image* (2005)—were important testimonies of emerging scholarship in the field across Indian universities around the turn of the century.[9] Privately owned satellite television started to increasingly figure in the agenda of television scholars from this time. More recently, Shanti Kumar's (2006) work on television and postcoloniality, Nalin Mehta (2008) and Daya K. Thussu's (2008) accounts of news television, Vamsee Juluri's (2003) work on Indian music television and Shoma Munshi's (2010) book on soap opera have used the lens of television to review issues of class, gender, citizenship, democracy, law, industry, labour, ethnicity, identity, and sexuality in the Indian context.[10]

The 12 chapters in this book examine a number of aspects of contemporary Indian television—news, soap operas, reality TV, television flow and televisual affect, Doordarshan, cable television, and regional programming—in order to offer a broad view of the current state of television and television studies. The chapters are ordered such that there is a thematic unity between consecutive pieces. The first four contributions—by Roy, Mankekar, Asthana, and Kumar—seek to set new directions in media scholarship by arguing for the adoption of fresh theoretical perspectives. The second group of authors—Sinha, Thussu, Mehta, and Hutnyk—are concerned with liberalization and its aftermath and focus mainly on television news in order to analyse the changing dynamics between information and entertainment. The chapters by Sen, Chakrabarti, Gupta, and Vangal examine various genres (reality TV, soaps, and news) in Indian television with special reference to the interfaces of the global with the national and the regional.

The call for new directions is issued in four distinct ways. In 'Television, Narrative Identity, and Social Imaginaries: A Hermeneutic Approach', Sanjay Asthana uses Paul Ricoeur's philosophical hermeneutics to look at three Doordarshan serials, *Maila Anchal* (1987–8), *Rag Darbari* (1986–7), and *Godan* (2004). Asthana scrutinizes the narrative structures of these adaptations to show how the pre-capitalist performative practices figure in televisual reconfigurations of myth and realism. While Asthana looks for the meanings embedded in formal structures, Purnima Mankekar turns to affect theory in order to recommend a turn away from questions of representation and signification to a consideration of the haptic and the corporeal. In 'Televisual Temporalities and the Affective Organization of Everyday Life', she analyses the 24×7 television coverage of the Mumbai attacks (November 2008) to point out how live television worked below the level of meaning—in the realm of affect—to generate a sense of time standing still. Television, she argues, not only represents and tells meaningful stories, it also has a capacity for engendering various levels of temporality and corresponding orders of affect. The pieces by Asthana and Mankekar exemplify how contemporary Western theory can be productively deployed to yield rich insights about categories like narration and temporality in Indian television. Roy and Kumar, on the other hand, insist that any valid analysis of Indian television (and media in general) has to be placed within the dual frames of postcoloniality and globalization. In 'TV after Television Studies: Recasting

Questions of Audiovisual Form', Abhijit Roy asks us to rethink Raymond Williams' notion of 'flow' by placing it in the context of globalized postcolonial societies. Roy questions the way television studies, due to a preoccupation with Western histories of the novel and the film, assume a certain correspondence between form and ideology to critique the notion of flow. According to him, forms historically corresponding to 'flow'—exhibiting a strong ideological framework despite a deceptive formal heterogeneity, that is, a dissonance between form and ideology—should be sought in contexts having a legacy of the colonial modern, like the Indian popular film form. Such formal correspondences between the flow form and an indigenous aesthetic crucially conditions the spectatorship and reception of television in India, and broadly in the non-West. A similar imperative is also at play in Shanti Kumar's chapter, 'Spaces of Television: Rethinking the Public/Private Divide in Postcolonial India', which sets out to critically evaluate the many definitions of the 'public' in Indian television in relation to the global trends in media privatization. Kumar draws on feminist and postcolonial theories to argue that binary category systems such as 'public' versus 'private' are highly problematic as they refuse to acknowledge the hybridity of such category systems in the culturally heterogeneous and ideologically overdetermined terrain of globalization.

Each of the subsequent four chapters in the book deals with the transition from state monopoly to a market-driven system comprised of a variety of local, national, and multinational channels. News programming is crucial to this new order. At the end of 2010, India had 122 news channels and among the top 20 news channels in the country, nine were regional-language channels.[11] Three of the chapters—by Mehta, Thussu, and Hutnyk—scrutinize specific aspects of news broadcasting in order to theorize the effects of liberalization over the past two decades. In 'When Live News was Too Dangerous: The Early History of Satellite TV in India', Nalin Mehta sets out to outline the early histories of the private networks like Zee, NDTV, and Asianet in order to identify the unique confluence of factors that led to the demise of the government's monopoly over television news. Mehta compares the introduction of satellite programming to two earlier transformations in the media system—the 'newspaper revolution' of the 1970s and the 'cassette revolution' of the 1980s—to argue that the availability of privately produced satellite television has meant that 'television has opened up avenues that previously did not exist and brought many more people into the public arena'.

The other two chapters in this group take a far more critical view of liberalization's impact on television news. Daya Thussu's chapter, 'Television News and an Indian Infotainment Sphere', investigates how commercial interest has led to compromising traditional news-gathering and reporting practices, inaugurating an era of 'infotainment'. Excessive marketization, according to Thussu, has made news increasingly shrill, bipartisan, and noisy. In the process, symbiotic relationships have developed between news and new forms of current affairs and factual entertainment genres such as reality TV, blurring the boundaries between news, documentary, and entertainment. John Hutnyk's chapter, 'NDTV 24×7 Remix: Mohammed Afzal Frame by Frame', goes furthest in criticizing new television journalism. Analysing an episode of NDTV's talk show, *The Big Fight*, that focused on the case of the 'terrorist' Afzal, Hutnyk contends that the current mode of television journalism is instrumental in conjoining state violence with an audience that is all too eager to define its own normality against 'terrorist' acts of deviance. This complicity in the legitimization of the state violence is made possible by our television screens. While Mehta, Thussu, and Hutnyk, all assume a radical disjuncture between Doordarshan and the post-liberalization era of Indian television, in 'From Clients to Consumers: The *Missing Citizens* among the Indian Television Audiences', Dipankar Sinha uses an audience-theoretic approach to argue that these seemingly disparate models have much in common. In his view, the Nehruvian vision of democratic communication turned out to be a case of mechanistic message-based indoctrination of audience. Sinha argues that the audience is no more empowered in the privatized era where a model of hyper-entertainment for 'consumers' ensures that the audiences function as mere 'viewers' and not as critical citizens.

An understanding of Indian television involves an analysis of the specificities of the levels of the local, the regional, the national, the transnational, and the global, as well as an examination of how these levels constitute each other in order to produce contemporary forms of televisual culture. The next four chapters in the book examine specific instances of this complex process. In '*Big Brother, Bigg Boss*: Reality Television as Global Form', Biswarup Sen looks at one well-known reality format—*Big Brother*—in order to shed light on the complicated relationship between the forces of globalization, national and local cultural formations, and the dictates of commercially driven entertainment. By looking at the manner in which reality formats are produced as exportable franchises, and also investigating their formal properties as aesthetic objects, Sen shows how the immense popularity of reality TV in India is conditioned by the various logics of globalization.

Santanu Chakrabarti's chapter, 'The Saffron Hues of Gender and Agency on Indian Television', examines the extraordinary popularity of a group of serials—*Kyunki Saas Bhi Kabhi Bahu Thi*; *Kahaani Ghar Ghar Kii*; and *Kasautii Zindagii Kay*—all aired on the STAR Plus channel between 2000 and 2008. These three family sagas—produced by the Mumbai-based Balaji Telefilms and known as K-serials since all their names start with 'K'—are representative of an era of soaps which locate women exclusively in the domestic sphere, in contrast to earlier shows. Chakrabarti argues that this regression to domestic has to be ideologically connected to the constructions of the woman and the family by the Hindu right in India. By demonstrating the problems of celebrating the K-serials as a site of women's agency and empowerment, Chakrabarti critiques any attempt to appropriate such female agency into any brand of feminism.

One of the consequences of the cable and satellite revolution has been a massive increase in the size and scope of regional television. Nilanjana Gupta's chapter, '*Sange Thakun*: Bangla News Channels and Media-citizenry', looks at the phenomenon of 24-hour news channels in West Bengal to argue that the kind of citizenry constructed by this sort of regional programming is quite distinct from the larger national construction of citizenship that is the imperative of the national news channels. Unlike the distant objective stance adopted by national newscasters, the discursive register of Bengali news channels reflects a unique 'structure of feeling' that has two predominant components: a distinct moral economy based on intense emotionality; and an intimate style. Continuing with the focus on increasing relevance of regional television in India, Uma Vangal's chapter, 'Tears, Talk, and Play: A Window to Gender and Sexuality on Tamil TV', examines soap operas and reality shows on Tamil television from the perspective of gender. While a lot of Tamil television continues to be regressive and reinforce traditional values, Vangal suggests that a wide range of programmes have become a window of opportunity for gender and sexuality to take main stage in the public minds and discourse.

In the past few years, many scholars have argued for an 'opening up' of television studies that would involve going beyond the narrow confines of media history and theory. Such an approach would locate television at the centre of social and political histories and analyse it from the perspectives of other disciplines. As Charlotte Brunsdon (2008: 134) puts it: 'In some senses, television is too important to be left to television scholars; but at the same time, television scholars can show something of how we might understand this importance'. Thus, we would want to inscribe television within a series of histories: '...of twentieth-century nation-states, of the electrification of the home, of serialized fiction, of technologies of the self, of international image industries' (Brunsdon 2008: 134).[12] The purpose of this kind of inquiry would not merely be to understand broad historical processes *through* television, but rather to reframe the existing debates on these historical processes by situating television as an indispensable player in them. It is our belief that a study of Indian television conducted along these lines would greatly deepen our understanding of a range of phenomena: citizenship in the post-liberalization era, consumerism, the public sphere, democracy, participatory cultures, technological convergence, urbanity, domesticity, privacy, vernacular cultures, sexuality, and many others. A number of crucial questions will need to be addressed: is television contributing towards creating a space for deliberative democracy or is it increasingly emerging as an appropriator of the functions of official institutions of democracy?

To what extent are traditional modes of power and control getting reoriented due to the contemporary operations of the media? What are the grounds and conditions of the kind of media ethics we are witnessing today in India? What is television's role in shaping the emergent forms of publicness in India? We hope that this book will provide the groundwork for answering such questions and thus help constitute a future research agenda for the Indian television scholar.

Notes

1. See Nalin Mehta, Chapter 7, page 171, in this book.

2. Available at http://www.cable-quest.in/pdfs/FICCI-KPMG_Report_ 2012.pdf (accessed on 22 April 2012).

3. Available at http://www.cable-quest.in/pdfs/FICCI-KPMG_Report_ 2012.pdf (accessed on 22 April 2012).

4. Available at http://www.censusindia.gov.in/2011census/hlo/Data% 20sheet/HLO%20All%20Indicators.pdf (accessed on 25 March 2012).

5. For a persuasive argument along this line, see Arvind Rajagopal (2011: 11–12).

6. For a detailed analysis and critique of such a historically produced distinction, see Dipesh Chakrabarty (1992).

7. Some such scholars are Nicholas Garnham, Peter Golding, Graham Murdock, Robert McChesney, Eileen Meehan, Vincent Mosco, Dan Schiller, and Janet Wasko.

8. The work of Manjunath Pendakur (1991) is notable in this regard. Also, see Vanita Kohli-Khandekar (2010) for a fact-filled account of all the media industries.

9. See *Journal of Arts and Ideas*, no. 32–3, 1999; *Journal of the Moving Image*, no. 4, 2005.

10. An exhaustive analysis of key works in the field of Indian television studies is impossible and unintended here. References to academic works by the chapters in this volume do present a vivid picture of the expanse and worth of studies in Indian television, another reason why we are not attempting a detailed survey here.

11. Available at http://www.business-standard.com/content/general_ pdf/032511_01.pdf (accessed on 11 September 2011).

12. Generally also, there is a great surge over the last few years of a discourse reminding us that we are presently at a point where the certi-

tude of the axiomatic of television studies seems to be dissolving or at least looks unstable. See, for instance, John Caughie's (2010) 'Mourning Television' essay in this regard.

References

Brunsdon, Charlotte. 2008. 'Is Television Studies History?', *Cinema Journal*, 47(3): 127–37.
Caughie, John. 2010. 'Mourning Television: The Other Screen', *Screen*, 51(4): 410–21.
Chakrabarty, Dipesh. 1992. 'Postcoloniality and the Artifice of History: Who Speaks for "Indian" Pasts?', *Representations*, 37(Winter): 1–26.
Chatterjee, P.C. 1991. *Broadcasting in India*. New Delhi: Sage Publications.
Curran, James and Myung-Jin Park (eds). 2000. *De-Westernizing Media Studies*. London: Routledge.
Dudrah, Rajinder. 2002. 'Zee TV-Europe and the Construction of a Pan-European South Asian Identity', *Contemporary South Asia*, 11(2): 163–82.
Gupta, Nilanjana. 1998. *Switching Channels: Ideologies of Television in India*. New Delhi: Oxford University Press.
Hall, Stuart. 1986. 'The Problem of Ideology: Marxism without Guarantees', *Journal of Communication Inquiry*, 10(2): 28–44.
Juluri, Vamsee. 2003. *Becoming a Global Audience: Longing and Belonging in Indian Music Television*. New York: Peter Lang Publishing.
Kohli-Khandekar, Vanita. 2010. *The Indian Media Business*, 3rd edition. New Delhi: Sage Publications.
Krishnan, Prabha and Anita Dighe. 1990. *Affirmation and Denial: Construction of Femininity on Indian Television*. New Delhi: Sage Publications.
Kumar, Shanti. 2003. 'Is There Anything Called Global Television Studies?', in Lisa Parks and Shanti Kumar (eds), *Planet TV: A Global Television Reader*, pp. 135–54. New York: New York University Press.
———. 2006. *Gandhi Meets Primetime: Globalization and Nationalism in Indian Television*. Urbana-Champaign: University of Illinois Press.
Mankekar, Purnima. 1999. *Screening Culture, Viewing Politics: An Ethnography of Television, Womanhood, and Nation in Postcolonial India*. Durham: Duke University Press Books.
Mehta, Nalin. 2008. *India on Television: How Satellite News Channels Have Changed the Way We Think and Act*. New Delhi: HarperCollins.

Mitra, Ananda. 1993. *Television and Popular Culture in India: A Study of the Mahabharat*. New Delhi: Sage Publications.

Munshi, Shoma. 2010. *Prime Time Soap Operas on Indian Television*. New Delhi: Routledge.

Ninan, Sevanti. 1995. *Through the Magic Window: Television and Change in India*. New Delhi: Penguin Books.

Pendakur, Manjunath. 1991. 'Political Economy of Indian Television: State, Class and Corporate Influence in India', in Gerald Sussmann and John A. Lent (eds), *Transnational Communications: Wiring the Third World*, pp. 234–62. Newbury Park, CA: Sage Publications.

Rajadhyaksha, Ashish. 1990. 'Beaming Messages to the Nation', *Journal of Arts and Ideas*, 19(May): 33–52.

Rajagopal, Arvind. (2001). *Politics after Television: Hindu Nationalism and the Reshaping of the Public in India*. Cambridge: Cambridge University Press.

———. 2011. 'Notes on Postcolonial Visual Culture', *BioScope: South Asian Screen Studies*, 2(1): 11–22.

Schiller, Herbert I. 1971. *Mass Communications and American Empire*. New York: Beacon Press.

Smythe, Dallas. 1981. *Dependency Road: Communications, Capitalism, Consciousness and Canada*. Norwood, NJ: Ablex Publishing.

Thussu, Daya K. 2008. *News as Entertainment: The Rise of Global Infotainment*. London: Sage Publications.

1

TV after Television Studies

Recasting Questions of Audiovisual Form

ABHIJIT ROY

New Debates

Reading the 'inside' of television, its screen, sound, images, and other formal elements, is usually unfashionable. It generates, without fail, resentment from perspectives that see the worth of television studies only 'outside the box' in audiences networks, industries, and elsewhere. The 'medium', in any case, has lost its relevance in the larger terrain of media and cultural studies, all the more so due to the tenor of ontological pursuit in many medium theories. However, the key trajectory of debate in this area around Raymond William's notion of 'flow' has never really waned, though has been designated as bearing a 'confusing inheritance' of a notion that is now merely of historical interest. In the demeanour of announcing a closure of the debate, John Corner expressed the hope that the idea of flow was not creating further problems for television theories.[1] Milly Buonanno, in a recent book, is critical of the idea for its claim to being a 'defining characteristic' of television and thinks that the phenomenon of flow (in the sense of *experiencing* television as a stream of images) is only a minor aspect of the gamut of viewing patterns (Buonanno 2008: 30–6).

But far from weakening, the critical interest in the original concept of flow has seen a surge over the last few years, due particularly to the onset of digital television or what is sometimes called, in the context of technological convergence, Television 2.0. The main feature is the viewer's ability to time-shift by recording preferred programmes, with the option, if one uses software like TiVo, of cutting out commercials and similar 'interruptions'. William Uricchio (2005: 232), invoking Williams' trans-Atlantic experience of encountering the flow form in a Miami hotel, shares his fascination with hotel television systems, the Marriott in Cambridge in this case, offering a taste of new television:

> ...my television offered such features as interactive messaging, account updates, nearly 40 films on demand, and Sony Playstation, in addition to cable television and both closed circuit and cable teletext systems. Simply turning on the television provoked a staggering array of decisions regarding language, services, and menu options, all of which had to be dealt with if one wanted to watch television in any of its forms.

The overarching assumption in most of the commentaries on digital television is that the notion of flow, in its original sense of an ideological formation, is utterly insufficient to understand the world of digital television that offers its viewers 'complete control of their own television use, freed from the tyranny of the television networks, their schedules and offers' (Moe 2005: 773). The standard discourse denouncing the efficacy of the notion of flow usually celebrates freedom and viewer control in the new television that promises to 'render channels superfluous, offer unlimited content on-demand and provide full interactivity' (Moe 2005: 773). The uneasiness with 'flow' here seems to be stemming from the concept's insistence on conceiving television as an ideological apparatus. Corner, for instance, assumes a postmodernist rationale of looking at the variety of fragments and corresponding viewing patterns, suspicious of any effort to theorize a contingent abstraction of the whole.[2]

Not all commentators, however, make the critique of flow synonymous with postmodernist celebration of difference or, in the more specific context of digital television, enhancement of viewer's choice. Uricchio refers to two major shifts in the form of the viewer–television interface and corresponding engagements with the notion of flow: from programming-centred (maximum control by broadcasters), to active audience-centred (use of remote control device), to adaptive agent-centred (interactive digital platform) formations. Uricchio (2005: 257) argues:

> As we have seen, generational clusters of television technology and cultural practice have each been bound up in particular power dynamics and discursive strategies. Thanks to Williams, the concept of flow, as a repository for thinking about changing strategies for content management, can also serve as a metaphor for our changing notions of ideology. Although its meaning is different, this metaphor remains vital to a critical understanding and evaluation of our interface with the television medium.

The interesting turnaround in the recent commentaries on flow, be it on digital television or not, is the tendency to relook at the possible ideological contours of television's audiovisual form. Kumkum Sangari, in her analysis of the televisual form in the context of disaster-time TV, proposes that televisual flow can decompose totalizing master narratives but is itself constituted within the force field of neoliberal globalization. Drawing our attention to 'conjunction', the predecessor of the notion of flow in Williams' writing, Sangari (2009) shows how unforeseen connections between programmes can emerge in a viewer configuration very much within the flow form. But what is more significant here is the agenda that Sangari (2009) proposes concerning the televisual form: 'The formal and fluid modalities of the televisual form often seem, definitionally and self-evidently, to foreclose the place of ideology. Is there a way to propel the formal properties of television back into the political arena, and reposition them as necessary abstractions of the social terrain?'

It is from this vantage point, where new questions on the relationship between form and ideology are being raised both by critics who read the audiovisual configuration of television as an ideological formation and by those who don't, that I look at the question of televisual form and try to locate that vis-à-vis the Indian experience. I think there is ample scope of interrogating the basic axiomatic behind some of the key positions in the debate, if one looks from a perspective that is almost non-existent in studies of televisual form, that of the postcolonial career of television. I would not consider the emerging digital television experience and the corresponding questioning of the very notion of 'flow' since, in India, more than 70 per cent of television homes watch television in sync with the broadcaster's schedule.[3] DTH (Direct to Home) service providers offer the facility of recording programmes, that is, reorganizing the scheduled flow, but the ultimate digital context of completely doing away with 'interruptions' (hence with flow in Williams' sense) through technologies like TiVo is still absent in India.

But more than the statistical account of usage of a certain technology, what is important is to recognize 'flow' as a historically specific form of television, as a 'dynamic trans-generational medium' suggested by Uricchio (2005: 236), whose formal, technological, and cultural contours have always been transient and unstable (Uricchio 2005: 234–6). Irrespective of the spread of direct-to-home technology, what still continue to significantly shape the televisual ideoscape in India are the superflow of programmes and 'breaks' across numerous channels, the individual channel flows (closest to Williams' category), and the viewer flows produced by remote control and other devices.[4] Generally also, in the far more advanced contexts of digital television use through metadata protocols, like Europe and the United States (US), efficacy of a reworked concept of flow has been argued for with much conviction. I want to quote a rather long passage from Moe (2005: 779):

> The options available to viewers are still subject to the offers of the television networks and distributors. Advanced EPGs [Electronic Programme Guides] still only present what the distributor allows delivered through the set-top box—which is in effect an important gateway. And despite numerous niche channels and on-demand offers, the relative homogeneity of television's superflow—the content and themes available to viewers—is still striking.

Television's very basic link to broadcasting, the element of a live 'streaming seriality', looks also unlikely to loose its charge in the face of digital distribution, as Hilmes (2008: 161) argues, 'I believe it unlikely that audiences will be willing to give up the dense, twisting, compelling, multi-layered and aurally ingenious serial texts that are broadcast television's cultural legacy to whatever comes next'. Moreover, a great amount of participatory culture and fan activity around television nowadays thrives on liveness, the simultaneity of the viewer and the broadcast flow: voting in *Idol* and other reality shows, various music and game shows that invite live phone calls, messages, liveblogging, etc.[5] One can possibly also cite news as a genre that carries with it the imperative of watching 'live', rather than watching recorded or on the web at a different moment.

All I am pointing towards is the continuing need to contextualize and scrutinize the notion of flow across histories of television–viewer interface. I think restricting the notion to Williams' original sense of a possible structure of programming sequence is limiting, as we should be more interested in the broader premise in Williams' theorization, that of the relationship between form and ideology, and the way this relationship calls for reworking the notion of flow. The new institutional contexts are to be found possibly in emerging industries that mediate distribution, a sector 'composed of metadata programmers and filtering technology (variously constructed as search engines and adaptive interfaces)' (Uricchio 2005: 249). The notion of flow still seems to be a useful entry point into any discussion of audiovisual configuration of television and the viewer's encounter with it. The question is whether invoking the hitherto unaddressed Indian (broadly the postcolonial) television into this discussion can reset the terms of what has been so far an exclusively Western debate on televisual form.

Alternative Modernity, Postcoloniality, and the Televisual Form

It seems that most of the critiques particularly have a problem with a sense of 'uninterrupted movement' or of 'smoothness' associated with the term 'flow'. But what we lose sight of is the relative autonomy Williams ascribes to the segments that create a great degree of heterogeneity in flow. Williams finds the possible formal legacy of such a structure in the West in the 'internal variation and at times miscellaneity' of certain significant traditions of the pre-modern popular: dramatic performances with musical interludes, the almanac, the chapbook, the magazine which was 'invented as a specific form in the early eighteenth century' and was 'designed as a miscellany'. All these seem to allude to the experience of early popular performative traditions that lie outside (or on the fringes of) the organized capitalist production of cultural artefact. In fact, he reminds us that the word 'programme' has its 'traditional bases in theatre and *music-hall*' (Williams, 1990: 88; emphasis added). On another occasion, he compares the television experience to a mixed baggage of segments from various genres, that of 'having read two plays, three newspapers, three or four magazines, on the same day that one has been to a variety show and a lecture and a football match' (Williams, 1990: 95). In brief, Williams' theorization of television refers to a heterogeneous structure in which the segments keep the formal difference from each other alive to generate the effect of miscellaneity and which, at the same time, is capable of conjuring a synthetic image. This double take on two modes of representation in the notion of flow, connected to parallel layers of modernity itself, is somewhat undermined by most of the commentators who always seem to get the sense of only modern realist narrative in the notion of 'flow'. This seems to be the reason why they emphasize discontinuity and segmentation, as if these are antithetical to flow.[6] Flow can simultaneously hold on to ideas of the segment and the whole of appearance and dissolution, the momentary and the structural.

Of all the early critiques, Jane Feuer's position seems particularly productive as it acknowledges such dualities, the fact that television is constituted by a dialectic of segmentation and flow. Her contention that Williams should more accurately be meaning 'segmentation without closure'[7] can possibly be used to refer to the traditions outside the modern realist narrative modes, that is, the vaudeville and variety shows in the Western context and the 'alternative modern' (Pinney 2004: 204)[8] forms in the non-Western context. The idea of a 'viewing strip', a serial set of programmes actually viewed as the unit of analysis proposed by Newcomb and Hirsch (1994: 509–10), also accommodates, along with the heterogeneity and intertextuality of programmes, the possibility of conceiving a sort of global image that constitutes the ideological location of the viewer. That the apparently loose and flexible structure of flow does not restrain elements of regularity was suggested very early by Heath and Skirrow (1977). They point out that the 'movement' of flow can only be conceived as a 'stasis' (Heath and Skirrow 1977: 15). This idea comes very close to that of the tableau mode of address in the early film form, with the all-important difference that here the ideological trope is not as loose and flexible as it could be in a pre-capitalist form. The 'stasis' refers as much to the consistency in the pattern of construction of subjects, as to 'flow', a certain arrangement of images and sounds that we come across whenever the TV is switched on.

Compared to the flurry of countercurrents in the theorization of televisual flow, intellectual inquiry into the mode of address, graphic organization of frame, and the nature of reception of television has produced a relatively consistent set of positions. The traits on which a reasonable degree of consensus appears to exist are: frontality of address; specific modes of reception embedded in family; the absence of the Western premise of realism in the disavowal of narrative continuity; capability of extracting intimacy and participation from the viewers; reduced sense of depth in frame; the currency of the extra-apparatus elements in constructing 'externality of spectatorship'; the open-endedness of segments; and so on.[9] One can add to these the variety show effect and the ideological 'stasis' of the televisual flow form that, as we have tried to show, bear the legacy of representations outside the framework of modern realist narrative. What is interesting to note is that the popular film form in India has been described in somewhat similar terms. Indian popular film form has been widely theorized as harbouring traits that are incompatible with the classical Hollywood film form: frontality of address, exhibitionism, discontinuity in narrative thread, the predominance of 'spectacles' as opposed to causal continuity, the condensed semantics of the iconic image, etc. These traits do not represent the autonomy of a set of unfiltered cultural/performative 'traditions', but they rather exist in film as vibrant conduits of negotiating the modern institutions of social transformation.[10]

Ravi Vasudevan's (2000: 138) comment relating to frontality in Indian film can effectively be applied to a broader understanding of recorded audiovisual performance:

> At one level frontality would mean placing the camera at a 180 degree plane to the figures and objects constitutive of filmic space. These may display attributes of direct address, as in the look of characters into the camera, but a frontal, direct address is relayed in other ways, as in the way the knowledge of the spectator is drawn upon in constructing the scene, through the stylized performance, ritual motifs and auditory address that arise from a host of Indian aesthetic and performance traditions.

The Indian popular film's investment in the iconic is perhaps the only trait that needs to be further explained in order to identify its formal correspondence to the televisual apparatus. According to Geeta Kapur (1993: 23), the iconic is 'an image into which symbolic meanings converge and in which moreover they achieve stasis'. It is our effort to theorize the flow-form television as a site that makes possible a renewed negotiation with the modern for the alternative modern subject, the key addressee of consumerist television. Thus, 'television' itself (whatever is the programme, we watch television, assuming the primary mode of viewing in India: encountering the broadcast flow) becomes a repository of aspirations for the viewer in such a context. One should remember here that it is only a set of formal association that, I am suggesting, exists between the Indian popular film and consumerist television. The corresponding tropes in ideology cannot be said to have the same function, though they can be historically related in interesting ways. Is it simply a coincidental analogy, merely a resonance in discourse that should not be overread, or is this correspondence capable of opening up unexpected junctures in the theorization of television?

Form and Ideology in Television

Before trying to account for this correspondence, let us get back to the flow debate for a while. As has been shown, most of the Western critiques have largely not accepted flow as an ideological formation on the ground that the form is heterogeneous, discontinuous, and open-ended. The assumption here is of a direct correspondence between form and ideology: if the form graphically exhibits qualities of homogeneity, continuity, and closure, there must be a corresponding coherent ideology; if the form engenders heterogeneity, rupture, and open-endedness, no overarching ideological agency can be envisaged. I think this sense of a direct correspondence between form and ideology emanates largely from a historically specific experience in the West: that of the rise of modern popular forms of cultural representation in certain correspondence with the modern capital. One can look not only at the classical film form that is read as having a direct correspondence with large-scale organized capitalist production of film, but also at West European novel in the nineteenth century associated with what is sometimes called print capitalism. The historical conditions of the 'classic realist text', are then seen as being absent in the corresponding pre or proto-modern forms like the early film which, hence, offers a rather scattered and heterogeneous structure. I argue that flow-form television cannot be deciphered in such terms of a direct correspondence between form and ideology, and this is precisely the point one can arrive at while trying to engage with Williams. Flow refers to a heterogeneous and discontinuous form, but may not be so at the level of ideology. It rather is a form that has *appropriated* the contradictory charges of formal heterogeneity itself, that has incorporated the devices of pre-modern discontinuity (early film) or modernist iconoclasm (Godard or new German cinema) into an absolutely novel formation in the history of audiovisual apparatuses. The Western critique's stance of rather *naturally* transferring the loosely strewn form to a corresponding looseness in ideology is a perception historically conditioned by a certain Western experience of the historical relationship between form and ideology. I suggest the examples of a strong ideological framework, despite a deceptive formal heterogeneity, should be sought in contexts with the legacy of colonial modern. The Hindi popular film form, in this sense, remains aptly comparable with the flow form. The difference, however, should not be missed: the former does not bear the logic of appropriation like the flow form, but is born out of a different historical encounter with the modern, discussion of which would be impossible in this premise. Let us concentrate only on the historical conditions that make the formal correspondence between flow and the Indian popular film possible.

Television seems to be a catalogue of devices that once aspired to contradict the terms of institutional modes of representation in film. In fact, the gradual rise in the reach and popularity of commercial television in America and Europe at about the same time as the demise of the classical Hollywood form is not at all coincidental. All the erstwhile devices of filmic iconoclasm found a new formal application, without their iconoclastic charge, in another apparatus at a moment when the cinematic institution was finding it difficult to sustain its bond with the 'modern', with the values of realism, continuity, and closure. In Europe, given the legacy of a realist film form, the emergence of the flow form would embody a major shift. I think the 'shock' that Raymond Williams exhibits in his description of the encounter with American television is indicative of the perspective of primarily the cinematic modern, or its somewhat corresponding form in public television (discrete programming, as Williams suggested).

The Indian Context

My suggestion is that India's encounter with 'flow' did not mark a shift in this sense. The shift here was posited rather by the state television, which tried to align itself to a realist–developmentalist aesthetic concomitant with a state-sponsored cinema.[11] It was a shift, not as much for the urban upper and middle classes who were relatively competent to grasp the codes of realism, as for the rural addressee who was expected to master the codes of what was a relatively new audiovisual regime for them. One major instance is the Satellite Instructional Television Experiment (SITE), undertaken by the Indian state during 1975–6. The state television, with an agenda of development not unconnected with its parallel coercive drives during the Emergency, failed considerably in enchanting the rural peasantry by its realist and documentary mode of address. While the failure of SITE can be attributed to a number of reasons, the discord between the preferred popular forms (nautanki, jatra, ramlila, various folk performative traditions, and above all, the Hindi popular films, for instance) and what Ninan (1995: 18) describes as 'dull programming', seems to be a key one.

The huge upsurge of films on Indian television in the satellite era cannot be explained simply by the dictates of 24-hours programming. The order of formal conformity between television and the Indian popular film is of particular significance here. Transmission of popular films on state television during the heydays of development communication in India was surely curtailed by the moralist citizenizing process that the broadcast media exhibited as part of the nationalist politics of development. While the popular cinematic public was conceived as one belonging primarily to the streets and lumpen public spaces, the territory of the broadcast public was essentially the home and the family, the sanctity of which, it was thought, could better sustain a citizenship premised upon the patriarchal order of the nation–family bondage. The broadcast public, directly under the aegis of postcolonial state pedagogy, could never be the libidinal 'vulgar' public of popular cinema, the constituency of capitalism's enchantment with the erotic, the vestige of colonial expansion worldwide of 'entertainment'.[12] But it is curious to note how a certain ideological pre-occupation curbed the way for a certain form to emerge on Indian television. The documentary-style realism associated with the discrete programming form highly suited the state television's investment in the Nehruvian socialist agendas of development and 'national integration'. In spite of being highly impoverished in resource, the state channel could not think of considerably exploiting the cheap option of telecasting films, as would be possible in the era of private satellite programming.

The moment television started expanding from 1982 through the 1990s, based primarily on the logic of market expansion, the apparatus currently in vogue started showing its signs: increased number of 'breaks', reduced duration of shots resulting in a speedier movement of the sequence. I am suggesting that this was far more resonant with the indigenous forms of entertainment connected to the legacies of an unresolved modern, to an incomplete process of bourgeoisification of the state form in India, to the simultaneous existence of a variety of production relations. By indigenous forms of entertainment, I do not mean a homogeneous set. I am referring rather to a popular pan-Indian aesthetic that has predominantly come to be embodied in the Hindi popular film form in India. While it is true that one needs to consider the evolution of this form, especially in terms of the emergence of particularly the super-genre of the 'Social' after Independence that subsumed various studio-era genres within itself, the 'variety show' look of the Hindi film was always actually present. Still, if one wishes to underline a particular period or form, it would be the 'all-inclusive Hindi film', the Social, that most aptly exemplifies the negotiations with the indigenous traditions of performance and reception. Most crucial here is to remember, as Madhav Prasad suggests, that the historical condition for such a negotiation is 'an unevenly developed market capitalism'. To Prasad (1998: 46–50), 'the ideology of the all-inclusive film, whose vision of the world tends to be multi-faceted, episodic and loosely structured,' should be related to a production process in which the logic of product differentiation, as in organized capitalism, 'has not advanced beyond the elementary stages'.

My conviction is that the televisual subject in India, and broadly in the non-West, continues to be historically related to these conditions of possibility of the Indian popular film form, keeping in mind of course that these conditions should be read as relatively autonomous, not essentially as ones predestined to dissolve in the wake of global capitalism. An abstraction of the 'popular' in the Indian context as the frontal, the spectacular, and the non-continuous, as representing the inclination of the tastes of the majority of people, of the sensibility of 'not-so-modern' subject, has been productive in understanding not only the popular film form in India but, broadly, a host of manifestations of the indigenous performative traditions. This 'popular' however, as

Pinney (2004) reminds us in his study of the 'visual history' of India, is best described as an 'alternative modernity' (see Note 8) and can only be theorized, not as the mark of an ontological value or a different historical location, but as very much a coordinate in the cartography of modernity. Ashish Rajadhyaksha, in his endeavour to locate the points of negotiation between the popular cultural forms and a nascent but struggling cinematic apparatus in India, also invokes the continuing efficacy, in shaping the realms of modern, of an identifiable set of pre-capitalist commercial representational forms in indigenous artefacts. In case of the negotiated colonial modern space of painting, and later film, in India, visuality is oriented more towards the 'iconic' and what has been called 'darsanic', both of which refer to a certain formal staticity and a gaze engaging in the vertical axis of reception, triggering 'identification' (Rajadhyaksha 1987). The point here is to recognize that the televisual flow is less proximate to the visual dynamics of classical narrative cinema with the latter's import of causality and resolution, than it is to the stasis of television as a whole. Flow, of course, refers to a dynamics, but of a sort that itself becomes a spectacle, an ever-present event that we want to play over and over again. The palpable orders of semblance can also be discerned in what Rick Altman has described as television's capacity to generate, through its overt investment in sound, a parallel 'household flow'.[13] While one can see the implication of this in the reception of soaps by the family at home, the significance of orality, music, and sound in the Indian, and broadly postcolonial, popular aesthetic in contrast to the ocular-centric modernity of the West is not difficult to locate here.[14]

We, however, do not want to be trapped in the teleological history of a sort of pre-figuration of the apparatus in the postcolonial and deny the specific histories of the emergence of the flow form in the West. The relationship can perhaps be rightly described as 'homologous' in the sense that Raymond Williams uses the word on another occasion: 'at one level correspondences are resemblances, in seemingly very different specific practices, which may be shown by analysis to be both direct and directly related expressions of and responses to a general social process'. He then goes on to chart various possible relationships from analogies to displaced connections, and finally concedes that the concept of homology indicates 'a sense of corresponding forms or structures' (Williams 1977: 101–7).

The sense of such correspondence of flow with a local popular is so insistent in the constitution of the televisual subject in this part of the world that denying its force becomes difficult. The homology would not be less operative in the relatively 'modern' subject, who, despite being armed with the legacy of classical/literary realism, is simultaneously vulnerable to the vibrant appeals of an indigenous aesthetic. I propose that this correspondence tends to orient the televisual subject in India in a grid of identification. It is not identification per se in the psychoanalytic sense of the term deployed in studies of classic realist film. It is rather a certain perception of 'us', not of 'I', that identifies its major expressive conduit in the televisual form here. The communitarian form of participation in the traditional entertainment performances possibly finds a resonance, with significant alterations, in a live interactive network of viewers. The drive for the televisual subject here is to *re-member* the 'self' over and over again into the televisual flow staging a network. The consumer subjectivity hitherto unaddressed by the state television would not only want to be acknowledged by the empowering agency of flow-form television, a key advocate of consumerism in a globalizing world, but would also want to see its preferred form, albeit in a negotiated manner, perpetually present in the apparatus (see Rajadhyaksha 2002).[15] To further elaborate on this, one can look at an very interesting connection between the moments of the birth of popular film and popular television in India. Both of these moments are tied by a highly suggestive commonality, the 'mythological' as the key genre of popularizing the medium. The mythological stories, being widely rooted in public knowledge, could be shown as a series of discontinuous episodic spectacles and fitted both the media's imperative to acknowledge a form with which the audience identified. The majoritarian (Hindu) ethos in the institutional construction of the popular at both of these moments triggered vibrant debates around 'identity politics' in India.[16]

The telecast of *Ramayan* and *Mahabharat* (1987–90), the Hindu epic serials, had as their conditions of possibility the 'flow', a form that sprang, not accidentally, affective modes of identification for a large constellation of viewers. In the section called 'how has television changed the context of politics in India?', Rajagopal (2001) significantly refers to a relentless attraction to television even when one is critical of the content. His acknowledgement of the role that 'national television' played in bringing split publics to the fore is indicative of various orders of subjectivity that consumerist satellite television can simultaneously bear (Rajagopal 2001: 25–6).[17] He refers to the way the Hindi newspapers represented the Ram Janmabhoomi movement in sympathetic terms as opposed to the set of English ones that had a more secular and objective tone.[18] To extend this very useful line of analysis, I would suggest that the huge reading public of the vernacular newspapers, who are by statistics far larger than that of the English newspapers in India, represent a certain subject that has a wider role to play in determining television's form, precisely because it is this subject's role in consumption that is most sought after by the drivers of the apparatus. The synchronic melange of a variety of components that commercial television has to nurture as an imperative would be particularly close to the subject less appropriated by unifying principles of modernity, and this can never mean taking the pleasure off the relatively 'modernized' subject. The latter, I argue, shall extract pleasures of representation along with identification. It is the television news and reportage that tend to generate in the relatively modernized subject, identification with 'flow' as the device of scanning and of analysis. It is to be remembered that such subjectivities are never exclusively embodied, but remain overlapped in the viewer. Attention to the formal layer of signification that constitutes the television apparatus and that has the capability of triggering intense viewer identification seems to further substantiate Rajagopal's meticulous analysis of an intense investment in *identity* at the moment of birth of popular television in India.

That the notion of flow is inconceivable without a particular context of competition-oriented market has already been pointed out. What remains to be said is that this market's drive to expand spatially to all the corners of the world in a certain period is somewhat proportionate to the intensity of the televisual flow.

The consumerization of a society cannot be unrelated to the extent of capital inflow that it has received from outside. In this sense, the transition from discrete programming to 'flow' not only marks a shift from a system of closed governmental control to that of a porous market, but also increased mobility in the flow of capital and a concurrent inflow of foreign satellite channels making the 'viewing strip' loose its discrete nature. Seen from this perspective, the flow form is the inevitable signifier of the global flow of capital and of images and narratives.

In brief, we are suggesting that the recognition of a body of formal devices in television in a postcolonial context is tied not only to a local legacy, but largely to a certain history of capitalism's renewed drive to expand spatially in the post-Fordist era. This however can wrongly be projected as a simplistic thesis of cultural imperialism, the subtle ploy of a global institution to generate strong identity through its apparatus in a prospective market. But the formal identity between flow and a certain Indian popular aesthetic refuses to be explained by the grid of ideological critique alone. At a certain juncture in history, in conjunction with the imperatives of an all-inclusive politics of consumerism, there emerged the need to incorporate the hitherto unaddressed 'lowly' forms into the dominant modes of cultural production in the West. Though this was manifest primarily in the thematic concerns, I propose that a certain section of audience can never subscribe to the emergent ethos of participation that television invites, unless the former, to a certain extent, *recognizes* the form. We shall remember here the reference to vaudeville and variety show that Williams made to characterize the commercial televisual form. The cultural logic of late capitalism cannot only be the thematic appropriation of subcultures and of the past of a certain class through nostalgia industry. It has to be helped by a certain graphic organization and mode of address that, to an extent, is continuous with the erstwhile cultures of the emergent consumer and that, especially in the mammoth terrain of the heteronomous popular, can never be only modern in the literary sense of the term. The early film form, which primarily entertained the working class in the American and European contexts in the early twentieth century, is indeed, to a reasonable extent, proximate to the colonial modern film form like India's. This is where I think the primary reason for the adjacency[19] of the flow form and the Indian popular film form lies: a drive to incorporate the residual cultural forms in the West and simultaneously to spread across the world, especially in the realms where the pre-modern is not marginal. These two drives are associative in their effort to incorporate the subject hitherto marginally addressed. This makes it possible for television to contingently choose, among many options, a form that can appropriate the residual and redefine the emergent, in conformity with the inclusionist rhetoric of consumerism.

In the context of the global flow of new capital along these lines, we shall now see how a specific type of mercantile culture emerged in India in the mid-1970s and helped generate the internal conditions conducive to the workings of the televisual apparatus. The parallel running of a developmental project and a state of Emergency (1975–6) should not only be seen as two mutually constitutive practices of the same political agenda, but primarily as the symptom of crisis of a certain state form. What we usually lose sight of is that the crisis was largely economic which the multinationals exploited by inserting cheap and pirated gadgets. Rajadhyaksha (1990: 37) points out that the 'Emergency in India was, in many ways, an effort to shore up, as almost a last-ditch stand, the orthodox protectionist measures of a national market in India—a promise that had been especially made to the Indian bourgeoisie during the freedom struggle' and that it was a sort of response to the 'flood of cheaply manufactured electronic durables'. The expansion of a capitalist regime had two primary effects that can account for the imminent rise of the televisual form as the key site of commercial aesthetics. First, the state itself tried to expand its grip through 'production control' and shift from the geographically controlled distribution systems to licensing over areas hitherto unattended, efforts that can not be isolated from SITE's effort to reach the rural India with the new technology of television. Second, this led to a surge of artisanal modes of production in small localities, the widespread availability of 'cheaply manufactured electronic durables using various modes of memory-recall and information storage—from the radio-and-tape recorder, two-in-ones to calculators, to various kinds of easily piratable computer software, and of course the VCR', which the state 'could do nothing to control' (Rajadhyaksha 1990: 37). This shows how the early phase of new global entrepreneurial drives gave rise to small artisanal sites and local modes of reception, relocating the erstwhile rural and small town popular into a frame of large communitarian dissemination. Of special interest here would be the rise of Santosh, the manufacturer of tape recorders and radios, and Super Cassettes Industries, the publishers of 'T-series' brand of music through audio cassettes, both being instrumental in popularizing the emergent modes of dissemination in the territory of the underprivileged and in exposing the regional cultural frame to the emergent market. These companies, for a long time, largely inhabited a grey market disavowing the modern dictates of copyright and exhibited pre-capitalist traits of production. T-series, that was possible in 1979 precisely because of the post-1975 situation I have tried to describe, 'started with a small studio where they recorded Garhwali, Punjabi and Bhojpuri songs...' (Manuel 1993: 67). The emergence of multiple desires for local dissemination of new global technologies should be taken as the condition of possibility for the identitarian grid of engagement with television, the new screen for global consumer culture offering semblance of a local popular form.

The writings on television in India have so far been indifferent to the debates concerning the flow form or the television apparatus in general, and have been interested more in televisual content's power to represent, reflect, and constitute subjectivities. The television apparatus should be read not merely as a technology for dissemination of content, but as a cultural form that orchestrates content and reformulates tradition and subjectivity in ways specifically conditioned by history. Our effort was not to devise an Indian theory of television, but to conjecture on a possible theory of television that can account for the particularities of a non-Western context.

Notes

1. See Milly Buonanno (2008: 31–2) for a discussion on Corner's and other such views. For an elaboration of Corner's position, one can see Corner (1999: 60–9).

2. See, in this regard, Rosemary White (2003).

3. According to an estimate in March 2013, DTH constitutes 30% of satellite (cable and DTH) homes in India which penetrate around 80% of Indian TV households. This means that the percentage of DTH homes in India is approximately 24%. See http://www.business-standard. com/article/companies/dth-readies-for-final-assault-on-cable-business-113100101283_1.html and http://cii.in/WebCMS/Upload/em%20 version%202_low%20res.PDF (both accessed on 26 December 2013).

4. Moe (2005: 778–9) cites Bruhn Jensen: 'He distinguished between three types: firstly, every television network or company plans a *channel flow* to keep the viewer watching for as long as possible. This is the category closest to Williams' original concept when describing textual organization on a macro-level. Secondly, every viewer creates her or his own *viewer flow* based on all presently available content. Here, the understanding of the subjective experience of each viewer, as also included in the original concept, is in focus. Finally, these two categories can be related to everything that is available on all channels—television's *super-flow...*'.

5. See http://mauramc.wordpress.com/2009/04/11/television-flow-and-liveness/ (accessed on 07 March 2010) for the perception of a liveblogger.

6. For a detailed analysis of critical reflections on the notion of flow, see Abhijit Roy (2005).

7. Jane Feuer, as quoted in Mimi White (2002).

8. Christopher Pinney (2004) prefers the term 'alternative modern' to 'pre-modern' to describe a set of north Indian visual cultural practices that lie outside the standard codes of 'modern' realist visual culture.

9. For an early discussion on these traits, see John Ellis (1992a, 1992b). Almost all books on television that have been published so far, more or less subscribe to these generalizations on television. I am mentioning only Ellis' book because it is one of the most influential books, first published in 1982, that undertook the task of characterizing television. The only characteristic that has drawn substantial contestation is the idea of 'glance' as an exclusive way of looking at the television. For a representative critique of this notion, see Buonanno (2008: 36–41), who feels that the television screen can also be a 'primary focus of interest, attention and gaze' (Buonanno 2008: 39), along with, of course, the glance as a way of looking.

10. For an analysis of the way these devices have been used to negotiate modernity in the context of the Hindi film, see M. Madhava Prasad (1998: 1–26). See also Ashish Rajadhyaksha (1987).

11. Ashish Rajadhyaksha rightly recalls the connection between the progressivist series like *Hum Log* and *Buniyaad* with the project of new Indian cinema of the 1970s. See Rajadhyaksha (1990: 41–2).

12. See Roy (2008) for a detailed analysis of the troubled relationship of Indian popular film with Indian public television and the All India Radio until the mid-1980s. The instance of banning Hindi popular film songs from radio by B.V. Keskar in 1952 perhaps can be best understood as the state's effort to curb the entry of this popular in the sacred confines of home/family that was supposed to be the exclusive constituency of state-controlled mass media. To telecast *Bobby* to disrupt an opposition rally during the Emergency (Ninan 1995: 26) is actually a curious testimony of the state's acknowledgement of the popular film as a mass entertainer, though a corrupting one. The same was reflected when film songs were brought back on radio with the official press release acknowledging that the film songs cost the broadcasting organization too much in popularity. In brief, public television always exhibited a certain degree of uneasiness with the popular film, particularly its 'commercial' aesthetics, thereby also curbing a certain form that was greatly preferred by the audience of development-oriented television in India.

13. Rick Altman, as quoted in Dienst, op.cit. pp. 29–30. The primacy of sound in television is also discussed extensively by Ellis (1992b).

14. See Abhijit Roy (2008) for a discussion on how a certain aurality becomes the key constituent of the postcolonial popular.

15. Rajadhyaksha hints that one of the reasons for film being so frontal in the Indian context is that here, popular cinema largely remains patronized by the people outside the empowering frames of modernity. This section, by presenting itself in the film theatre, exercises the modern democratic rights that it is usually barred from doing outside the theatre and derives a certain sense of being 'empowered' out of it. 'We can now start seeing why the cinema came to mean so much to Indian people at large. Those basic "enumerative" rights of democracy—the right to be counted, the right to receive welfare—are precious, since these are often the only kinds of rights that people in general basically have in a place like India, and people to different degrees are aware of and "recognize this"' (Rajadhyaksha 2002: 283). '...At this level, therefore, when the viewer purchases a ticket, enters the auditorium and "releases" the film, saying "I am here" ("I am present... I help it to be born"), what the cinema is also doing is to incarnate one of the most fundamental, even if ambiguous at times, rights of democracy. The inscribed viewer category comes in precisely through the invitation to viewers to identify it, and then identify with it: hence, both statements, "It is true" and "Do you recognize what you see?"; or even, in admittedly different kinds of political contexts, "You see this/You own this". What is to be "recognized" is not necessarily what is on screen: what is also recognized is that the abstract category of viewer whom the film is, so to

Channeling Cultures

say, "supposed to" be addressing..."' (Rajadhyaksha 2002: 283). I think, television spectatorship, though inconceivable in terms of the exercise of democratic rights in a public place like film theatre, is nevertheless crucially premised in the disempowered viewer in the non-West. The viewer, by entering into a parallel frame of democracy, surveying the market, and choosing the right television set, and ultimately 'owning' the TV set, and hence subscribing to a certain publicly accepted order and daily routine at home, would demand a form that would enable a transaction between the modernizing institution and the existent popular of the actual viewer and would recognize 'the abstract category of viewer' whom the television is 'supposed to be addressing'. The moment of television is a moment when strong identitarian attachment is sought from those who had been hitherto considered relatively insignificant in the spread of capitalism. But more than deciphering the apparatus in terms of merely the class, a geopolitical understanding can be helpful. Television is also capable of invoking strong identificatory engagement in the upper-class viewer of the Third World as well, due primarily to television's reassuring gesture that the much-despised subservient position of the postcolonial bourgeois vis-à-vis their Western counterpart can now be erased. The point is that though the natures of dissemination of film and television are different, the theorization of flow-form television cannot ignore the way historical encounter with film—the other audiovisual apparatus—in India has been theorized.

16. See Rajadhyaksha (1987) for the implications of Gandhi's invocation of the category of 'swadeshi' to refer to primarily the Hindu mythological films. For understanding the intricacies of 'identity politics' around the telecast of *Ramayan* and *Mahabharat* in Doordarshan and Hindutva's claim to the state form, see Arvind Rajagopal (2001).

17. Rajagopal (2001: 152) talks about the '"boundary-piercing character of television"', while referring to television's capability of enacting the conflict between the erstwhile split publics. I would suggest that along with staging antagonism, television can create a broader level of semiotic exchange between, and coexistence of various subjectivities and identities and thus become a quasi-nation, quasi-globe space. And precisely because identities tend to increasingly collate, chances of extra-televisionic conflicts escalate. The peculiar relation of popular television to the intensification of identity politics in India, in this sense, also involves questions of televisual form.

18. 'A "split public" in the making and unmaking of the Ram Janmabhoomi movement' (Rajagopal 2001: 151–211).

19. Paul Bove uses Edward Said's term 'adjacency' to refer to a certain lineage of the poststructuralist approach to 'systems of thought'. Tracing

a major source of influence over Foucault in Georges Canguilhem's contention that 'different sciences and systems of thought "cohere"', Bove suggests that this approach forces to consider '...how, within the "systems of thought" they constituted, various "sciences" might be institutionally and even conceptually discontinuous; how they might be practiced, as it were, at disparate points within a culture and yet, given their "adjacencies", make up a coherent system of thought spread across a range of institutions and discourses whose family resemblances can be traced by the genealogist interested in their multiple origins, transformations, and their value for the present' (Bove 1990: 55–6).

References

Bove, Paul. 1990. 'Discourse', in Frank Lentricchia and Thomas McLaughlin (eds), *Critical Terms for Literary Study*, pp. 50–65. Chicago and London: University of Chicago Press.

Buonanno, Milly. 2008. *The Age of Television: Experiences and Theories*. Bristol, UK: Intellect Books.

Corner, J. 1999. *Critical Ideas in Television Studies*. Oxford: Clarendon Press.

Ellis, John. 1992a. 'Broadcast TV as Cultural Form', in John Ellis, *Visible Fictions*, pp. 111–26. London: Routledge.

———. 1992b. 'Broadcast TV as Sound and Image', in John Ellis, *Visible Fictions*, pp. 127–44. London: Routledge.

Heath, Stephen and Gillian Skirrow. 1977. 'Television: A World in Action', *Screen*, 18(2): 7–59.

Hilmes, Michele. 2008. 'Television Sound: Why the Silence?', *Music, Sound, and the Moving Image*, 2(2): 153–61.

Kapur, Geeta. 1993. 'Revelation and Doubt: Sant Tukaram and Devi', in T. Niranjana, P. Sudhir, and V. Dhareswar (eds), *Interrogating Modernity: Culture and Colonialism in India*, pp. 19–46. Calcutta: Seagull.

Manuel, Peter. 1993. *Cassette Culture: Popular Music and Technology in North India*. London and Chicago: University of Chicago Press.

Moe, Hallvard. 2005. 'Television, Digitalisation and Flow: Questioning the Promises of Viewer Control', available at http://cicr.blanquerna.url.edu/2005/Abstracts/PDFsComunicacions/vol2/08/MOE_Hallvard.pdf (accessed on 12 August 2011).

Newcomb, Horace and Paul Hirsch. 1994. 'Television as a Cultural Forum', in Horace Newcomb (ed.), *Television: The Critical View*, 5th edition, pp. 503–15. New York: Oxford University Press.

Ninan, Sevanti. 1995. *Through the Magic Window: Television and Change in India*. New Delhi: Penguin Books.

Pinney, Christopher. 2004. *Photos of the Gods: The Printed Image and Political Struggle in India*. New Delhi: Oxford University Press.

Prasad, M. Madhava. 1998. *Ideology of the Hindi Film: A Historical Construction*. New Delhi: Oxford University Press.

Rajadhyaksha, Ashish. 1987. 'The Phalke Era: Conflict of Traditional Form and Modern Technology', *Journal of Arts and Ideas*, 14–15: 47–78.

———. 1990. 'Beaming Messages to the Nation', *Journal of Arts and Ideas*, 19(May): 33–52.

———. 2002. 'Viewership and Democracy in Indian Cinema', in Ravi Vasudevan (ed.), *Making Meaning in Indian Cinema*, pp. 270–85. New Delhi: Oxford University Press.

Rajagopal, Arvind. 2001. *Politics after Television: Hindu Nationalism and the Reshaping of the Public in India*. Cambridge: Cambridge University Press.

Roy, Abhijit. 2005. 'The Apparatus and its Constituencies: Notes on India's Encounters with Television', *Journal of the Moving Image*, 4, available at http://jmionline.org/film_journal/jmi_04/article_01. php (accessed on 12 August 2011).

———. 2008. 'Bringing up TV: Popular Culture and the Developmental Modern in India', *Journal of South Asian Popular Culture*, 6(1): 29–43.

Sangari, Kumkum. 2009. 'Conjunction and Flow: The Gendered Temporalities of (Media) Disasters', *Journal of the Moving Image*, 8, available at http://jmionline.org/film_journal/jmi_08/article_06. php (accessed on 7 September 2011).

Uricchio, William. 2005. 'Television's Next Generation: Technology/ Interface Culture/Flow', in Lynn Spigel and Jan Olsson (eds), *Television after TV: Essays on a Medium in Transition*, pp. 232–61. Durham, NC: Duke University Press.

Vasudevan, Ravi. 2000. 'The Politics of Cultural Address in a "Transitional" Cinema: A Case Study of Indian Popular Cinema', in Christine Gledhill and Linda Williams (eds), *Reinventing Film Studies*, pp. 131–64. New York: Arnold.

White, Mimi. 2002. 'Flows and Other Close Encounters with Television', Paper presented at the conference entitled 'Media in Transition 2: Globalization and Convergence', at the Massachusetts Institute of Technology, 10–12 May, available at http://web.mit.edu/cms/ Events/mit2/Abstracts/MimiWhite.pdf (accessed on 17 March 2010).

White, Rosemary. 2003. 'Review of John Corner's book, *Critical Ideas in Television Studies*, Oxford: Oxford University Press, 1999', in *Film-*

Philosophy, 7(15), July, available at http://www.film-philosophy. com/vol7-2003/n15white (accessed on 6 April 2011).

Williams, Raymond. 1977. 'Typification and Homology', in Raymond Williams, *Marxism and Literature*, 101–7. Oxford University Press.

——— (edited by Ederyn Williams). 1990. *Television: Technology and Cultural Form*. London: Routledge.

2

Televisual Temporalities and the Affective Organization of Everyday Life

Purnima Mankekar

My objective in this chapter is to propose a theoretical framework for analysing the role of television in the temporal and affective organization of everyday life. Hence, rather than embark on an empirical investigation of the 'impact' of television, I am interested in raising some *conceptual* questions about how television, in particular satellite television produced for audiences in India and beyond, participates in the generation of regimes of affect and temporality. Transnational satellite television networks have frequently been analysed in terms of space, as, for instance, in discussions of satellite television's spatial spread across the world. Indeed, as implicit in terms like flow and dissemination, spatiality is the dominant trope in scholarship on media in cultural studies, as evident in paradigms like Stuart Hall's (1980) classic model of encoding/decoding, and more generally, in conceptions of the circuits and circulation of translocal media. Yet, we forget that all circulation also occurs *in time* and *over time*, and that temporality is as fundamental to the work of media as is spatiality.

My curiosity about televisual temporalities first arose when I watched the reiteration of the images of the attacks on the Twin Towers of the World Trade Center in New York on 11 September 2001. Few of us can forget those images: they became fodder for all of us concerned with the role of television in the organization of everyday life and of the production of crisis and national emergency. My interest in television and temporality intensified when I watched the coverage of '26/11' on satellite television. Following an older preoccupation, namely, the role of television in the production of nationalist affect (Mankekar 1999), I was struck by how televisual representations of catastrophe and national emergency play a critical role in the affective organization of everyday life.

As I argue in my forthcoming book, media like television play a crucial role in the conjunction of affect and temporality. My conception of affect draws on scholarship on structure of feelings (Williams 1978), the anthropology of emotions and sentiments by Catherine Lutz and Lila Abu-Lughod (1990), Michelle Rosaldo (1984), and Sylvia Yanagisako (2002), as well as my own earlier explorations of the role of television in the construction of the affective bases of nationalism (Mankekar 1999); it also draws on theorizations of affect by Baruch Spinoza (1985), Sara Ahmed (2004), and Kathleen Stewart (2007). Lutz and Abu-Lughod (1990) have argued that emotions cannot be tied to tropes of interiority and are, therefore, not exclusively located within the domains of the psychic, the cognitive, or the subjective. As I conceptualize it, while emotion and affect are not coterminous, affect is also not tied to tropes of interiority. Affect cannot be located solely in an individual subject, nor can it be relegated to the psyche or to subjective feelings. Subjects are not where affect originates; rather, affect forms subjects through the traces it leaves upon them. My conception of affect draws on the formulations of Deleuze (1990, 1991) and Massumi (2002) who insist that affect is distinct from feeling (the domain of individual subjectivity) and emotion (the domain of the social and the linguistic). Affect is often experienced in the body, yet cannot be biologized. Thus conceptualized, affect alerts us to bodily modes of engagement. While most analyses of television have centred on the gaze and the visual, it is helpful to also consider the role of the haptic (as in cooking shows, travel shows, or many kinds of leisure programming) and the corporeal (as when our bodies register anger, fear, or erotic arousal as a result of our engagement with television). Analysing affect hence enables us to engage corporeal dimensions of our engagement without biologizing them.

Similar to a structure of feeling that cuts across individual feelings and collective sentiment (Williams 1978), affect transects, and therefore blurs, the binary between private feelings and public sentiments (see, for instance, Berlant 2008 and Stewart 2007). At the same time, in contrast to the work of sociologists like Emile Durkheim (1966, 1976) who located sentiments such as anomie or collective solidarity resolutely within the social, I argue that affect is irreducible to the social. Yet, affect is profoundly refracted by sociality. Like a structure of feeling, affect is not reducible to ideology. Yet, ideologies gain considerable political force through the ways in which they acquire affective freight: consider, for instance, the affective potency of nationalist ideologies.

In this chapter, I outline the role of television in the production of specific affective regimes and explore how these affective regimes implicate temporality. I use the term affective regimes for the following reasons. First, I wish to shift our focus from individual responses or experiences to how affects might slide across subjects and, in so doing, *constitute* subjects. Second, I wish to foreground the imbrication of affects with socio-political institutions such as the nation-state, the market, the patriarchal family, and so on. Finally, although engaging an ethnographic study of how viewers interpreted specific television programmes is beyond the scope of this chapter, in using the term affective regime, I wish to stress that affects are never totalizing in their effects. Audiences of television comprise of not just textual subjects but historical and socio-political subjects whose responses to the affective regimes generated by television are refracted by a host of factors, including their positions vis-à-vis axes of gender, class, caste, and religious affiliation.

For someone whose politics (and career) has centred on feminist critiques of nationalism, watching the televisual coverage of

26/11 via satellite at home in Los Angeles made for an admittedly intense affective experience. As I watched, my grief about the loss of life became subsumed by my growing discomfort generated by the jingoism of news reporters and commentators on what was unfolding before my eyes. Yet, without a doubt, the minute-by-minute coverage kept me riveted: life, literally, came to standstill for me as I cast aside all professional and social responsibilities to watch what was happening. My engagement with the coverage of these events in Mumbai in November 2008 compelled me to rethink how television shapes the experience of temporality, and how it imbues temporality with a range of affect.

Catastrophe, crisis, emergency are, of course, fecund sites for such an analysis. But I also want to go beyond televisual representations of purportedly singular episodes or events such as catastrophe, crisis, and emergency to see how television shapes a range of temporalities, ranging from the banal and the everyday, to notions of historicity, contemporaneity, and futurity.

Twenty-four-Hour News and the Temporalities of Crises

In examining the modes of affective engagement demanded and constructed by different televisual genres, I intend to produce neither a survey of Indian satellite television nor an ethnography of viewers' interpretations of specific programmes. Instead, as noted earlier, my goal is a lot more modest: I am interested in raising some broader theoretical questions about the role of television in the construction of temporality and the affective organization of everyday life. Many analysts have noted that the live 24-hour coverage of the violence in Gujarat pogroms was a watershed in the history of Indian television, particularly in its representation of the violence as a national emergency. Indian television's coverage (and production) of '26/11' was also very much about liveness, about getting a minute-by-minute account of what was happening. Unlike the Gujarat violence when television reporters were able to infiltrate the very landscapes of violence, the 26/11 crisis was constructed through footage of spaces outside the disaster-struck areas. There has been a great deal of debate on the relationship between the spaces occupied by the media, especially television reporters, and the creation of a sense of national in/ security at that time. But if the coverage of the Gujarat violence resulted in a sense of spatio-temporal expansion—we were 'led into' the scenes of violence—the coverage of 26/11 resulted in spatio-temporal compression. The fact that reporters could not follow the story' but had to be fixed in space, in front of the Taj Mahal Hotel or outside the Chhatrapati Shivaji train station, waiting for the story to unfold, was, I posit, crucial to the production of affective regimes of anxiety and fear that resulted from the act of waiting. Live television, in this instance, was about waiting for events to unfold, generating a sense of time standing still.

In this instance, the temporalities of national emergency were profoundly imbricated with the affect of anxious waiting, and were critical to the hegemonic production of 26/11 as a distinct event analogous with '9/11'. Many scholars have problematized representations of 9/11 as a 'turning point', 'watershed', or 'rupture', as a singular event that was unprecedented and exceptional (see, for instance, Dudziak 2003). This sense of emergency, of exceptionality (cf. Agamben 2005), was then deployed and manipulated to condone draconian attacks on civil rights within the United States (US) and human rights elsewhere. In a fashion similar to news media in the US, Indian news channels, such as Headlines Today, Aaj Tak, STAR News, and Zee TV, represented 26/11 as exceptional and as a watershed. Many crises produce an interruption of everyday life but, in the case of live television coverage of 26/11, our sense of interruption was compounded by a simultaneous feeling of urgency, of time ticking by—especially for the hostages. Although a formal analysis of the coverage of the 26/11 violence lies outside the scope of this chapter, it seems clear that a sense of interruption was built into the narrative logic of the reportage of these events as crisis and emergency. Put another way, the temporality of liveness was co-implicated with the sense of the interruption of the everyday: all other programming, including advertisements, was suspended. The presence of anchors and reporters at the scene was, of course, oriented to giving us live, 'breaking' news, to transporting us to the scene of the action;

the point was that we were supposed to be (presumably) living it with them minute-by-minute through the affective and temporalizing regimes generated by the television coverage.

What kinds of affective regimes were generated by the chronotope of live coverage of catastrophe? The close conjunctions of spatiality with temporality produced specific affective regimes. The reiteration of particular kinds of footage, for instance, the shots of Hemant Karkare, the police officer who was allegedly killed by terrorists on 26 November 2008, in the act of wearing his bullet-proof vest, the very vest that turned out to be inadequate to protect him, was framed through the production of affective regimes of grief at his martyrdom and admiration at his courage. Similar affective regimes were produced by the reiteration of portrayals of the body of Major Sandeep Unnikrishnan, the army officer who was killed in the terrorist attacks on the Taj Mahal Hotel in Mumbai. The affects generated by these shots were imbued with a distinct temporality of mourning: an interruption of the rhythms of everyday life was suffused with an acute sense of time standing still.

In most instances, the temporality of death is distinct because it interrupts everyday life even as it constitutes it. But in the case of the live coverage of 26/11, the distinct temporality of death was made all the more particular by how it was kept alive on television. In the coverage of 26/11 (and this is true of the live coverage of many scenes of catastrophe and death, including 9/11 and the tsunami of 2003), reiteration also served to keep the presence of death alive—and here we see the distinct temporality produced by live television. As Doane (1990: 238) remarks, 'The death associated with catastrophe ensures that television is felt as an immediate collision with the real in all its intractability.'

Affect, Temporality, and Everyday Life

It is important to note that televisual temporalities do not emanate only from the coverage of catastrophe. Consider how television produces more banal aspects of the everyday through traffic reports or weather reports. These, too, are important sites of the construction of liveness as realness. Liveness is sometimes deemed to distinguish television from cinema (see White's [2004: 76] critique of this assumption). The temporality of liveness is crucial to the formation of realness—what Feuer (1983) terms as the ontology of liveness as ideology. The hegemonic implications of this constitution of liveness, its politics, are obvious. But it is also important to remind ourselves that liveness is not distinctive of television; the Internet has now become an important source of getting the news 'as it breaks', as is the radio (few South Asians of my generation could have missed the excitement of live cricket commentary on the radio!). Further, it is crucial that we acknowledge the heterogeneous modes of engagement entailed in 'watching' television these days, for instance, through prerecorded programmes on TiVo or on digital video recorders (DVRs), video cassettes and DVDs of popular television programmes, the Internet, and increasingly, on mobile phones. Temporality is also structured and engendered through the modality of repetition, and this is evident not only in the repetition of headlines on 24-hour news channels where images and copy are reiterated but also through reruns of successful programmes. How do these technological and stylistic developments articulate with the temporality of liveness and contemporaneity? For instance, what happens when you can watch a particular episode hours, or days, or weeks after it has aired, or re-watch it? How do these viewings shape our experience of temporality?

In recent years, reality television has become immensely popular in India (major successes include *Foodistan* [NDTV Good Times; January 2012]; *Indian Idol* [Sony Entertainment Television; October 2004–current]; and *Sacch ka Saamna* [STAR Plus and Life OK; July 2009–April 2012]). These programmes raise important conceptual and empirical questions regarding the relationship of the ontology of reality shows with the ontology of 'live' programming, with profound implications for how time is experienced. However, televisual temporalities extend beyond live or, indeed, reality programming. It is critical that we note the different temporalities produced through diverse modes of address found on television, all of which are imbricated with a range of affective regimes. Diverse temporalities are at play in the ways in which television structures the everyday for its viewers. The role of episodic narration in shaping temporality is particularly salient in the serialized narratives that have been popular in India since at least the mid-1980s. I am reminded of Tania Modleski's (1983) now-classic work on the everyday rhythms of soap operas in the US (1983), but I would also like us to think of how the temporality of the Hindi television serial in which the continuity (and interruptions) of narrative, and the diegesis itself, is embedded in the very logic of the everyday. Consider, also, the marking of time through the modality of predictability of daily television programming: a day that begins with breakfast news, morning programmes (frequently consisting of reruns, which have their own distinct temporality) and talk shows, of movies in the afternoons, and news and lifestyle programmes in the evenings. Thus, while it is important to note that the recent phenomenon of narrowcasting might have scrambled this particular patterning of the organization of everyday life, most viewers' experience of the everyday is often mediated by the predictability of television programming.

Affect, Temporality, and Historical Consciousness

How does television mediate the relationship between the experience of the everyday and the sedimentation of memory and historical consciousness? In this regard, it is crucial that we raise some conceptual questions regarding televisual mediations of *histoire* and *duree*, and its reconfiguration in conjunction with the time–space compressions of satellite television. Let us begin by considering the role of television in the formation of historical consciousness. This happens, first of all, through programmes that claim to be historical (The History Channel being the most obvious), but it also occurs in all documentaries and films that claim to narrate 'the past'. But television mediates the formation of historical consciousness also through its cumulative effects, the concatenation of the everyday into the experience of the present; for instance, when the fleeting ontology of the daily news crystallizes into the sense of the contemporary. John Caldwell's (1995: 28) formulation of the 'all-at-once-ness' of global television might well apply across different genres: certainly the all-at-once-ness of television feels particularly acute on MTV-like programmes and on news programmes where multiple news stories are narrated simultaneously. These programmes entail different modes of engagement: we listen to the news anchor even as our eyes drift across the different ticker tapes, sometimes across split screens.

Consider, also, how the *duree* is affectively produced through our sense of contemporaneity. Despite their apparent evanescence, news programmes are an excellent example of how television produces the duree in terms of contemporaneity. The temporalizing processes of news television may be banal, or momentous and world changing. News programmes might consolidate the tedium of the everyday even as they enable forms of witnessing what is unfolding on the 'national' or 'global stage'. News programmes are both performative (in that they enact the now-ness of contemporaneity) as well as pedagogical (in teaching us how to experience the passage of time). The immediacy of news programmes produces heterogeneous affective investments, for instance, rage, fear, or feelings of triumph. Recall the centrality of Indian news channels to representations of the long-drawn-out battle between Sri Lankan security forces and the Liberation Tigers of Tamil Eelam (LTTE) in 2009. In southern India, where I was living at the time, the affective regimes generated when audiences witnessed the brutal repression of the LTTE by the Sri Lankan army were complex and heterogeneous: their reactions ranged from ambivalence to frustration and a sense of defeat, and varied sharply according to their affective investment in the Eelam struggle. The affective regimes generated by television coverage of the conflict and its repression were much more predictable in northern India: there, the publics created by television reacted with relief at what they perceived as the defeat of 'terrorism' in neighbouring Sri Lanka. Clearly, the work of television in the construction of public meanings is deeply imbricated with its role in the construction of public sentiments or communities of affect,[1] as well as its mediation of how the passage of time, in this case, the time of conflict and civil war, was experienced. The coverage of the defeat of the LTTE at the hands of the Sri Lankan army suggested how television can configure our experiences of temporality through its ordering of events in particular sequences so as to not only imbue them with causality but, perhaps more importantly, invoke and produce a range of affective regimes.

Analysing the work of television enables us to unpack the relationship between the temporality of histoire and the duree, between historical consciousness and the everyday. Questions of history also bring to the fore the role of Indian television in the blurring of myth and history or, rather, the transformation of myth into history. There have been many examples of this, most notably serialized epics like the *Ramayan* (Doordarshan; January 1987–July 1988) and *Mahabharat* (Doordarshan; October 1988–June 1990), which have played a central role in the production of affectively-charged representations of a glorious Hindu past: this blurring of the line between myth and history has been critical to the consolidation of Hindu nationalism. Consider, also, the reiteration of images of the body of Indira Gandhi lying in state after her assassination in terms of the centrality of television to the construction of memory and memorialization. The complexities of the production of history through memory were also brought home to me when I interviewed viewers of the television serial *Tamas* (Doordarshan, 1986) for a previous project. As I found, wilful forgetting and silence are as foundational to historical consciousness as are memory and memorialization.[2]

In addition to television's role in the blurring of myth and history, however, let us think of the narrative logic undergirding television's mediation of the relationship between the duree and histoire. I would like to revert to my earlier example of television soap operas, many of which go beyond the consolidation of notions of the everyday and, consequently, mediate the affective production of a sense of duree for viewers. The one that comes immediately to my mind is *Kyunki Saas Bhi Kabhi Bahu Thi* (Because the mother-in-law was once a daughter-in-law) which received very high ratings when it was first launched in 2000 by STAR Plus. At the time, pundits within the industry and in academia attributed its success to the network's dexterity in indigenizing (Indianizing) its programming for audiences in India: *Kyunki* was viewed as an exemplar of 'localizing' global programming.

Here, again, we see the dominance of tropes of spatiality to the neglect of the temporalizing processes that were to play a critical role in the establishment of an entire genre of television programming, the *saas–bahu* (mother-in-law/daughter-in-law) genre of television serials which continues to be ubiquitous on Indian television channels. As implicit in the very title, *Kyunki Saas Bhi Kabhi Bahu Thi*, the blurring (yet distinctness) of the roles of mothers-in-law and daughters-in-law suggests a particular temporality of cyclicality and, hence, of continuity. It is the patriarchal extended family and the relationship of mothers-in-law with daughters-in-law, represented as cyclical and as fundamentally unchanging, that undergird the *purported* constancy of a social institution which is posited as the last bastion of tradition against the onslaught of global modernity. These representations of the patriarchal family as under threat, yet ultimately resilient, have been the dominant motif of most of the saas–bahu serials that spilled over from the 1990s into the early years of the twenty-first century.

Much has been written about the conservatism of these serials and at the level of their plots, they are certainly conservative. I would like to argue, however, that conservatism of these saas–bahu serials was imbricated in the temporalizing processes that they generated. These constructions of the past *as part of* the present suggest how retrospection (the temporality of the past) and contemporaneity (the temporality of the present) are brought together. *Kyunki...*, as well as other serials of this genre such as *Kahaani Ghar Ghar Ki* produced a distinct temporality in terms of how they facilitated the collapse—or rather, the return—of the past into the present. It is no accident that the clothes, the sets, the interior landscapes of most of these serials made them seem so much like period dramas. They, as much as the serialized religious epics described earlier, encouraged the mourning (and through this mourning, the resurrection) of a particular kind of past through its representations of the patriarchal extended family as foundational to discourses of Indian culture. Put another way, these serials produced notions of traditionality that were often cast in terms of 'the vanishing' (Ivy 1995), as under threat by an increasingly invasive present troped as modernity.

And yet, let us not for a moment imagine that these representations of tradition are predicated on a rejection of all forms of modernity. The most cursory glance at Indian television's history from the 1980s onwards reveals its centrality to representations of aspects of modernity, in particular capitalistic modernity, as something *to aspire to* for the future. And thus, television has been foundational to discourses of futurity after the launch of policies of economic liberalization. Hence, for instance, there was an intimate relationship between the launch of entertainment serials on television from *Hum Log* onwards and the consolidation of middle-class practices of consumerism (Mankekar 1999; see also Rajagopal 2008). The consumerist practices associated with the launch of television serials in the mid-1980s and through the 1990s underscore how, for particular social segments, discourses of futurity are often cast in terms of aspirations, resulting in the production of a new demographic, 'the aspirational classes'. In another article (Mankekar 2013), I analyse, at great length, the commodity affects produced by advertisements, television serials, and music video. Here, suffice it to say that by yoking futurity to aspirations to consume, advertisements, home shopping networks, and product placement in television programmes have played a crucial role in the affective investment of many middle-class viewers in discourses of national progress that are co-implicated with incitements to consumerism.

* * *

In the larger work titled 'The Practice of Television in Everyday Life' from which this chapter has been extracted, I trace the place of spatiality in the cultural work of television; in this chapter, however, my objective has been to outline a conceptual framework that may allow us to examine the role of television in the production of multiple temporalities and affective regimes. My thinking of televisual temporalities is far from unprecedented: Raymond Williams' (1974) notion of flow remains one of the most important articulations of the imbrications of spatiality with temporality. My interest in this chapter has been in how we may theorize the participation of television in processes of world making through its multiple temporalities and its affective organization of everyday life.

Rather than exceptionalize television, it seems to make more sense to see how television articulates with other media that saturate our lives. Put another way, we need to shift our focus from a single medium, whether television, or cinema, or the Internet, by examining the semiotic and affective ecologies constructed by a range of media. At the same time, perhaps it is these multiple temporalities that enable us to see more clearly the distinctive power of television in a context where our socialities are constructed through a saturation of diverse media. Feuer (1983) has argued that this capacity for liveness is what enables television to 'insinuate itself into our lives' (White 2004: 81). There is no doubt that television enables the rearticulation of relationships of intimacy and proximity on an ongoing basis. Yet, exceptionalizing television might be particularly problematic in light of how new media are embedded in our lives, our socialities, and subjectivities, to the extent that they have compelled us to redefine what it means to be human. Similarly, we can hardly ignore the passionate intimacies fostered by cinema, the complex temporal registers enabled by the newspaper, and so on. Nevertheless, it is its 'multiple' temporalities that enable television to mediate the affective organization of the everyday—the conjunction of the everyday with the singular, the episodic with the continuous, the banal with the momentous—and it is perhaps this remix of temporalities that contributes to some of its power as a medium.

Notes

1. See Berlant (2008) for a broader discussion of public sentiments and communities of affect.

2. See Mankekar (1999) for an extended discussion of forgetting, silence, and historical consciousness pertaining to the Partition and anti-Sikh violence in 1984.

References

Agamben, Giorgio (trans. by Kevin Attell). 2005. *State of Exception.* Chicago: Chicago University Press.

Ahmed, Sara. 2004. *The Cultural Politics of Emotion*. New York: Routledge.

Berlant, Lauren. 2008. *The Female Complaint: The Unfinished Business of Sentimentality in American Culture*. Durham: Duke University Press.

Caldwell, John. 1995. *Televisuality: Style, Crisis, and Authority in American Television*. Brunswick, NJ: Rutgers University Press.

Deleuze, Gilles (trans. by Mark Lester with Charles Stivale). 1990. *The Logic of Sense*. New York: Columbia University Press.

———. 1991. 'Postscript on the Societies of Control', *October*, 59 (Winter): 3–7.

Doane, Mary Ann. 1990. 'Information, Crisis, Catastrophe', in Patricia Mellencamp (ed.), *Logics of Television: Essays in Cultural Criticism*, pp. 222–39. Bloomington: Indian University Press.

Dudziak, Mary L. (ed.). 2003. *September 11 in History: A Watershed Moment?* Durham: Duke University Press.

Durkheim, Emile (trans. by S.A. Solovoy and J.H. Mueller). 1966. *Rules of Sociological Method*. New York: The Free Press.

——— (trans. by J.W. Swain). 1976. *The Elementary Forms of Religious Life*. London: George Allen and Unwin.

Feuer, Jane. 1983. 'The Concept of Live Television: Ontology as Ideology', in E. Ann Kaplan (ed.), *Regarding Television*, pp. 12–22. Frederick, MD: University Publications of American Film Institute.

Hall, Stuart. 1980. 'Encoding/Decoding', in Stuart Hall, D. Hobson, A. Lowe, and P. Willis (eds), *Culture, Media, Language*, pp. 128–38. London: Hutchinson.

Ivy, Marilyn. 1995. *Discourses of the Vanishing*. Chicago: University of Chicago Press.

Lutz, Catherine and Lila Abu-Lughod (eds). 1990. *Language and the Cultural Politics of Emotion*. Cambridge: Cambridge University Press.

Mankekar, Purnima. 1999. *Screening Culture, Viewing Politics: An Ethnography of Television, Nation, and Womanhood in Postcolonial India*. Durham: Duke University Press.

———. 2013. 'Dangerous Desires: Erotics, Public Culture, and Identity in Late Twentieth Century India' in Purnima Mankekar and Louisa Schein (eds) *Media, Erotics, and Transnational Asia*. Durham: Duke University Press.

Massumi, Brian. 2002. *Parables for the Virtual: Movement, Affect, Sensation*. Durham: Duke University Press.

Mellencamp, Patricia. 1990. TV Time and Catastrophe, in *Logics of Television: Essays in Cultural Criticism*, ed. Bloomington: Indian University Press, pp. 240–66.

Modleski, Tania. 1983. 'The Rhythms of Reception: Daytime Television and Women's Work', in E. Ann Kaplan (ed.), *Regarding Television*, pp. 67–75. Frederick, MD: University Publications of American Film Institute.

Rajagopal, Arvind. 2008. *Politics after Television: Hindu Nationalism and the Reshaping of the Public in India*. Cambridge: Cambridge University Press.

Rosaldo, Michelle. 1984. 'Towards an Anthropology of Self and Feeling', in Richard Shweder and Robert LeVine (eds), *Culture Theory: Essays on Mind, Self, and Emotion*, pp. 137–57. Cambridge: Cambridge University Press.

Spinoza, Baruch (trans. by Edwin M. Curley). 1985. *The Ethics: The Collected Works of Spinoza*, Vol. 1. Princeton, NJ: Princeton University Press.

Stewart, Kathleen. 2007. *Ordinary Affects*. Durham: Duke University Press.

White, Mimi. 2004. 'The Attractions of Television: Reconsidering Liveness', in Nick Couldry and Anna McCarthy (eds), *Mediaspace: Place, Scale, and Culture in a Media Age*, pp. 75–92. London: Routledge.

Williams, Raymond. 1974. *Television: Technology and Cultural Form*. London: Fontana.

———. 1978. *Marxism and Literature*. New York: Routledge.

Yanagisako, Sylvia. 2002. *Producing Culture and Capital: Family Firms in Italy*. Princeton, NJ: Princeton University Press.

3

Television, Narrative Identity, and Social Imaginaries

A Hermeneutic Approach

SANJAY ASTHANA

It has been over half-a-century since television was instituted in India. Although scholarship on Indian television has provided valuable insights into its structural, textual, and reception contexts, a sustained examination of its narrative forms, for the most part, remains unexplored. In addition, a significant number of television serials translated from literary works have not been studied. In this chapter, I argue that a critical study of television's narrative forms, that looks at how its repertoire of cinematic codes, techniques, and styles involve a complex overlapping between myth, realism, and melodrama, will enable a better theoretical understanding of Indian television. In what follows, I discuss the television serials, *Maila Anchal* (1987–8, directed by Ashok Talwar), *Rag Darbari* (1986–7, directed by Krishna Raghav) and *Godan* (2004, directed by Gulzar), transmitted on the state-run television network, Doordarshan. These serials, adapted from literary works, deal with questions of late colonial and postcolonial identity in the countryside in terms of the dialectic tensions between the forces of tradition and modernity, local politics, nationalism, development, encroaching capitalism, public life, and more specifically, on the human condition. They frame the various issues through complex 'mythic-realist' narrative forms. For instance, the literary adaptations look at the effects of the process of development—through local (rural) politics, state institutions and bureaucracies—in the lives of ordinary Indians, mostly peasants and working class. Both a critique and valorization of anti-colonial and postcolonial nationalism, the television serials portray the stories and plots in terms of a complex interplay between the religious beliefs and practices with the secular vocabularies of the public institutions. The main focus of this chapter is to explore the 'ways' through which the literary works (novels) are translated for Doordarshan and the presence of pre-capitalist performative practices in the televisual reconfigurations of myth and realism.

To develop such an analysis, I engage with Paul Ricoeur's philosophical hermeneutics, and certain strands of neo-Marxist and postcolonial perspectives.[1] Ricoeur's work, largely ignored in cultural and media studies, offers productive avenues in examining multiple temporalities in television narratives in terms of the stories, plots, and the intersubjective webs of relationships between and among the characters and their life-worlds.[2] In the British context, a discussion on television and narrative forms, on the question of realism and melodrama, largely shaped by the film studies-based *screen theory* perspective, focused on the processes of televisual identification, narrative space, strategies of gaze, etc. On the one hand, this discussion, influenced by a reading of Marxist and Freudian formulations, had opened up a significant area of inquiry, and on the other, it also pointed to several theoretical conundrums in analysing television. Although such studies might consider the narrative structures, semiotic codes, psychoanalytic identification, narrative voice, etc., I argue that Ricoeur's hermeneutic approach enables a better analytic grasp into the televisual construction of myth, realism, and melodrama. More specifically, this chapter brings Ricoeur's concepts of narrative identity, temporality, and imaginary into a productive dialogue with postcolonial scholarship.

To pursue the connections between television and narrative form, I invoke Raymond Williams' (1977) work on realism that sought to provide a historical and contextual understanding of how realism is indeed a mediation of various cultural forms. If realism, following Williams, were to be understood as a highly variable and multi-textured concept, inflected by specific articulations of myth and melodrama, then the notion of development realism would be one among several other kinds of realisms. Scholars have pointed out that a slew of serials on the state-run television network, Doordarshan, particularly during the years 1983–91, articulated aspects of postcolonial Indian state's developmental ideologies embodied in ideas of family planning, education, progress, modernization, etc. Indeed, with respect to cinema, the postcolonial Indian state had earlier mobilized several filmmakers to produce films depicting state's development ideologies. Some of these filmmakers were later enlisted to produce television programmes for Doordarshan. While a majority of programmes represented state's development realism, a significant number of television series—*Charitraheen, Darpan, Godan, Kayar, Kathasagar, Kirdar, Maila Anchal, Mujrim Hazir, Rag Darbari, Neem Ka Ped,* and *Panchlight*—and single adapted plays—*Jazeere* and *Pita*—based on literary works from various Indian languages and the West, part of Doordarshan's programming, complicate our understanding of televisual narrative forms in terms of realism alone, and indeed open up the possibility of productively engaging with what Bisnupriya Ghosh (2004) noted as the significance of the vernacular literatures and idioms in consolidating postcolonial epistemologies.[3]

A study of television programmes based on literary works—novels, short stories, and plays—offers insights into how these are adapted for television, raising questions of aesthetic and formal techniques that underpin the transformation of 'form' from the literary to the audiovisual. The 'forms of television' discussed by Williams (1974) demonstrate how television combines various cultural forms like theater, newspaper, magazine, radio and cinema, and at the same time, alters these to fit the televisual modalities.[4] By incorporating the formal and aesthetic elements of various cultural forms, television changes the graphic organization of these forms; however, the communicational modalities of earlier forms remain salient. Thus, television analysis must begin with a consideration of how, and in what specific ways, the aesthetic and formal elements shape the forms. Williams's work, situated within the Marxist materialist approach, is not a call to a formalist analysis; rather, it is a first step towards examining the form–content dialectic in television.[5] Indeed, such an examination is similar to what Fredric Jameson (1981: 98–9) had outlined—through his third horizon of analysis of individual text or cultural artefact—as the 'ideology of form', wherein 'form' is apprehended as content.

Jameson indicated that 'the study of ideology of form is no doubt grounded on a technical and formalistic analysis in a narrower sense, even though, unlike much traditional formal analysis, it seeks to reveal the active presence within the text of a number of discontinuous and heterogeneous formal processes'(1981: 98). More importantly, in the Indian context, Jameson's notion, 'ideology of form', enables us to grasp how the literary and televisual text contains 'the dynamics of sign systems of several distinct modes of production'—particularly the imbrication of pre-capitalist, mythic, and popular elements with the capitalist ones in television series (Jameson 1981: 98). While it is useful to study how the different 'modes of production' coexist in the text, it would be more productive to examine, following Ricoeur, how, and in what particular ways, narrative, temporality, and social imaginaries reveal about human action at the level of narrative.

Discussions of television in India have been framed in terms of the singular concept of 'development realist' aesthetic underpinning state-initiated broadcast policies and programming. In the Indian context, the origins of several television forms can be traced to the late colonial period, 1920–47, when British administrators and bureaucrats sought to create radio formats and schedules for broadcasting. For instance, colonial broadcasting's centralized control, programme formats, and construction of audiences through essentialized cultural categories were in tandem with postcolonial state's upper caste–class Hindu ideologies. Overall, several postcolonial television programme formats—based on the development realist aesthetic—followed the ideas espoused by

Hardinge (1934: 619), a colonial bureaucrat, who pointed out that the 'peasant needs short daily talks of a homely nature upon rudiments of hygiene, sanitation, child welfare, improved agricultural methods and marketing, and similar helpful subjects leavened with entertainment'.

Since the 1970s, several programme formats of Doordarshan incorporated British colonial notions of rural development and the United States (US) brand of modernization. This is evident in the one-year Satellite Instructional Television Experiment (SITE) and the Kheda rural communication project—designed by India's technocratic–bureaucratic elite, in consultation with United Nations Educational, Scientific and Cultural Organization (UNESCO), and a satellite loaned from the US—where programmes incorporated developmental ideologies within pre-capitalist narrative forms. An important argument for starting SITE was based on the idea of television as enabling a 'participatory democracy' through several short skits and docudramas that mixed everyday issues with larger goals of family welfare, health, sanitation, social ills, etc., and aimed at rural students, farmers, and women. However, the developmental ideologies of SITE reproduced the colonial notions of 'village uplift' through a pedagogic and moralizing stance that assumed that such programmes could lead to an attitude change among people. The concepts of participatory democracy and attitude change, however, were unable to dislodge deeply entrenched feudal, caste, and class relations in the countryside.

The Doordarshan-initiated *Hum Log* (We people, 1984–5) and *Buniyaad* (Foundation, 1985–6), designed and produced as the televisual serial mode of composition and touted as 'progressive melodramas', marked an important shift from the state's development realist aesthetic, even as it embodied colonial and postcolonial developmental ideologies.[6] Indeed, as Roy (2008: 38) noted, 'one of the major sites of negotiation between the developmental State's pedagogic project and the emergent commercial popular was the "progressive melodrama"'. The arrival of commercial programming on Indian television in the 1980s, and the subsequent rise of satellite television in 1990s, led to the spread of a whole range of consumerist-based programme formats.

Channeling Cultures

In addition to the various programme formats discussed earlier, television in India has incorporated the formal language of commercial Indian popular cinema in developing specific televisual forms that adapt the cinematic techniques in programme production or directly draw upon the content of the films—as in song-and-dance sequence, cinema scenes, talk shows, etc. The borrowing and reworking of film music on television and the increasing pastiche of film-based television programmes may point to television's constant thirst for content.[7] Madhava Prasad (1997: 127) noted that 'apart from the readymade forms of the sitcom, the soap opera—the Indian television has borrowed from elsewhere, there are signs in some quarters of formal innovations that can be traced to the specificity of Indian film history'. Unlike in Britain and elsewhere, Doordarshan has been unable to draw upon the various Indian theatrical and dramatic aesthetics in developing hybrid televisual forms. While Veena Das (1995), among others, noted that aspects of aesthetic and pre-capitalist elements from Indian dramatic traditions have found their way into several television programmes, they have not shaped the television's narrative forms in any significant way.

Developing a series of interconnected arguments on the ideological features of television and nation, Roy (2005) persuasively argued that studies of television in India have not paid attention to the debates concerning television 'form.' Roy situates his analysis by drawing out the idea of 'flow' in Williams' work. The question of how various popular performative practices and traditions—from the pre-capitalist to the capitalist—have influenced television 'form' in India is suggestive. Indeed, as Roy points out, a few studies of Indian films offered original insights into aesthetic and pre-capitalist traits in Indian film form. However, what is more interesting to the purpose of this chapter is Roy's attempt to engage the materialist contexts of the televisual apparatus in India. For instance, he notes, 'the television form is replete with traces of capital affecting the production process, that it can hold the ideologies of both the state and the market in such a manner as to make each one of them capable of *representing* the other, can we say that the televisual form does incline towards an ideology of real subsumption?' (Roy 2005: 9).

Overall, I agree with Roy's argument that television can hold the ideologies of the state and the market, but would add that there are other ideological elements that may or may not fall under the sway of capital or state power. However, certain narrative forms of television are capable of either disengaging or reorganizing the dialectic of state and market to produce complex ideologies. Indeed, television series transmitted on Doordarshan developed such a complex set of ideologies ranging from conservative to radical. And as Williams (1977) and Brian Longhurst (1987) indicated, there are different kinds of realisms that televisual texts are capable of articulating.[8] Therefore, the mythic and realist elements in the television series, identified earlier, can be apprehended as narrative forms in terms of the aesthetic, formal, and political levels. Furthermore, as Kapur (1987: 106) noted, the issue of realism 'must be continued to be debated, especially in the third world, where realism as a genre has proved to be so hospitable, spawning all manner of realisms that replicate and reify the "given"'.

While studies of television soap operas and serials, and the overlapping of realism and melodrama, reveal interesting aspects of television forms, a hermeneutic exploration of television narrative forms informed by Paul Ricoeur's concepts sheds light on how television narratives are constituted; thereby providing a better grasp of the overlapping of aesthetic, formal, and structural modes of graphic and visual composition. In contrast to the various textual and narrative approaches that use semiotic and psychoanalytic concepts, Ricoeur's analytic posits four dimensions—the formal, the historical, the phenomenological, and the hermeneutic—and provides a comprehensive framework for examining a variety of literary and cultural texts, ranging from an 'almost infinite variety of narrative expressions (oral, written, graphic, gestural) and the narrative classes (myths, folklores, legends, novels, epics, tragedies, drama, films, comic strips, etc.)' (Valdes 1991: 27). Ricoeur's formal and historical dimensions enable a better grasp of the aesthetic and technical aspects through which television narratives are constructed.[9] On the one hand, they accord importance to the various technical and semiotic codes that structure the televisual narrative, and on the other, retain the important distinction between the semiotic and semantic levels.

Pursuing the key questions underlying hermeneutics, 'what is the meaning of being human', and how does a 'being/subject' come to interpret itself, Ricoeur posited that the meaning of 'being/subject' is mediated through the process of interpretations—cultural, religious, political, historical, and scientific (Kearney, 2004). Through his numerous critiques of structural linguistics, semiotics, and psychoanalytic approaches to language and subjectivity, Ricoeur argued that philosophers need to pay attention to the 'fullness of language'—since symbols, myths, metaphors, and even words constituting language carry traces of history and experience that cannot be revealed through structural analysis of texts via lexical codes and deep structures. In contrast to the structuralists, Ricoeur developed a theory of language through his philosophical hermeneutics by maintaining a distinction between semiotics and semantics: semiotics with its elaborate signifying system built on lexical codes and binaries, and semantics, where the sentence lives, and words accrete new meanings, thereby acknowledges the speaking/listening subject with his/her history. Referring to a 1967 essay of Ricoeur, where he presented his critique of the structuralism and outlined his approach to language, Andrew (2000: 58) noted that:

His [Ricoeur's] brilliant and characteristic move in this essay was to interpose a term between the dyad 'langue/parole' of Saussurian linguistics; that term, 'mot,' carries thick traces of theology and history, complicating what he saw as too simple a distinction. Every word, Ricoeur points out, bears in its etymology the sediment of prior uses that amount to a history of experience. History can be accounted for neither by structural rules (langue) nor by an accumulation of individual events (parole). Words—les mots—especially in their evolution, are what bear tradition, heritage, and the credit human beings can draw on for a shared future. Structural analysis of texts may be indispensable to an explanation of their power to make meaning, but it is completely inadequate to the task of comprehending their import and consequence.

Following Emile Benveniste, Ricoeur argued that meanings constituted beyond the sentence can be grasped at the semantic level, since it is here the subject of utterance and subject of enunciation 'live', and where words and sentences return to history bringing polysemy to language. Subsequently, Ricoeur began to examine the semantic plane and developed original insights into the nature of language and meaning, interpretation, subjectivity, and human action. It is in this context, he brings his philosophical hermeneutics in dialogue and debate with other disciplines.[10]

A significant feature of Ricoeur's work has been its ability to engage with diverse philosophical movements; however, as the Marxist theorist Jameson (2009) noted in his critique of *Time and Narrative*, Ricoeur ignored the Marxist tradition in his discussion of historiography and historical narrative.[11] While admiring the ambition of Ricoeur's *Time and Narrative*, Jameson (2009: 487) notes: 'What is bold in the ambition of the work itself is not only the vindication of narrative as a primary instance of the human mind, but also the equally daring conception of temporality itself as a construction, and a construction achieved by narrative itself'. Jameson's critique of Ricoeur is wide-ranging, and touches upon its many strengths and weaknesses. As Jameson (2009: 494) points out, 'despite his brilliant appropriation of Heidegger's complex temporal analyses, Ricoeur is apparently unwilling to entertain any possibility that human time has in late capitalism undergone a kind of structural mutation'. While Ricoeur's notion of multiple temporalities does not account for late capitalism and postmodern moment, it does provide significant analytic possibilities in examining heterogeneous times, at the level of history, narrative, and selfhood.

The Hindi novels, *Maila Anchal*, written by Phanishwarnath Renu in 1954, *Rag Darbari*, written by Srilal Shukla in 1968, and *Godan*, written by Munshi Premchand in 1936, are popular literary fictions that draw upon nationalist and historical topics, characters, and events from late colonial and postcolonial period. The multi-textured literary realism of the novels structured by their respective aesthetic forms and narrative strategies both subvert and expand the self-evident logics of realism to understand colonial domination and postcolonial power. In their literary and fictional writing, Premchand, Renu, and Shukla refigure and 'emplot' historical characters and events through the narrative. While it is well known that literary fiction makes explicit use of narrative, there has been a controversy among historians about the role of narrative in writing history.[12] Ricoeur (1984) provided a powerful argument that history writing takes the form of narrative, and that historians draw upon the techniques of literary fiction in ordering and conveying their arguments and ideas. Maintaining the distinction between 'fact' and 'fiction'—especially the truth claims about what constitutes 'fact' and 'fiction'—Ricoeur brings back the crucial role of 'plot' in historical and fictional narrative discourses. In the three novels and their televisual adaptations under discussion, therefore, the refiguration of myths and tradition is not just an ideological exercise, rather a creative and critical interpretation of aesthetic forms at the level of narrative.

Focusing on the impoverished conditions of rural landscapes and villages, the failure of the 1920s–30s anti-colonial nationalism (as in *Godan*) or the 1950s–60s postcolonial nation-state (as in *Maila Anchal* and *Rag Darbari*), the novels portray the impact of local and national politics on the life-worlds of the peasants. The novels *Rag Darbari* and *Maila Anchal*, narrated from multiple character positions, deal with the quotidian life in rural villages of India. Combining the mythic-realist forms and mixing genres, the novels present evocative blend of local politics and culture, the pervasive presence of state institutions and bureaucracy with the developmental and nationalist discourses. In *Rag Darbari*, *Maila Anchal*, and *Godan*, the actions are centred in the village with the arrival of 'outsider'—from the big town or city—as an important part of the story. While in *Godan*—organized around a 'social-realist' aesthetic that presents a searing critique of capitalism and hence is of a different order from *Rag Darbari* and *Maila Anchal*—the arrival of Gobar, Hori's son, is at the end of the story, *Rag Darbari* and *Maila Anchal* begin with the arrival of Ranganath, nephew of Vaidyaji, and Dr Prashant.

In his Marxist reading of Premchand's text, Michael Sprinker (1989) notes that Premchand reproduces the agrarian structures prevalent in northern India during the 1930s period and develops an accurate narrative of how the larger socio-economic forces lead to the plight of poor peasants. Pointing to the Marxist orientation of Premchand's narrative, Sprinker (1989: 65) adds, 'Premchand's understanding of the elementary fact of social life is registered more immediately in the text's overwhelming concentration on the intricacies of finance, which dominate the action and occupy most of the mental exertions of nearly all the characters, from the poorest like Hori and his family to the wealthiest like the Rai Sahib and Khanna'. In Sprinker's reading, *Godan* is understood as signalling a transition from feudalism to capitalism, and is considered as 'an authentically naturalist novel'. There is more to Premchand's literary form than Sprinker suggests. For instance, in the early 1930s, in a conversation with the Hindi writer Jainendra, Premchand delineated his literary style in terms of a 'complex' realism: one that is attentive to its own deficiencies and those of idealism and naturalism.[13] Explaining the development of Premchand's literary writings, Sudhir Chandra (1982) indicated that the delineation of nationalism in Premchand could be tracked in terms of the historical events surrounding anti-colonial nationalism.[14] In his analysis, Jagdish Lal Dawar (1996) identified a Gramscian strategy in Premchand's work, particularly through the critique of 'common sense'. For Premchand developed this 'from within the oppressed groups without being inculcated by the nationalist intelligentsia' (Dawar 1996: 117). There is no educated leader organizing the peasants: Hori, the main protagonist and the peasant encumbered by huge debt attributes the reasons for injustice and social inequalities in terms of actions preformed in previous life, unlike Gobar, his son, who provides a critique to Hori's common sense.

Renu's *Maila Anchal* and Shukla's *Rag Darbari* provide an evocative demonstration of how language operates on the semantic level, where words, as Andrew (2000: 58) stated, 'carry thick traces of theology and history', thereby enabling sentences to 'return to history', and where language and form conjugate in the development of narrative form. Indeed, as Ricoeur (1968: 126) argued,

> words are the point of articulation of semiology and semantics, in every speech event. The sentence, we have seen, is an event: as such, its actuality is transitory, passing, vanishing. But the word survives the sentence. As displaceable entity, it survives the transitory instance of discourse and holds itself available for new uses. Thus, heavy with a new use-value—as minute as this may be—it returns to the system. And in returning to the system, it gives it a history.

Consequently, Ricoeur's hermeneutic philosophy identifies the polysemic nature of language, while at the same time acknowledging the presence of heterogeneous time. In the following pages, I explore Renu and Shukla's work through Ricoeur's hermeneutic approach.

Renu's *Maila Anchal*, considered the most significant literary work after Premchand's *Godan*, is set in a north-eastern rural region of Bihar. Although Renu spoke the Maithili language of the region, and was conversant in Khari Boli, Bengali, and local dialects, he wrote primarily in Hindi. Kathryn Hansen (1981) noted that Renu developed a particular style of Hindi language based on rural speech and idioms and grafted words from one language to another (Sanskrit, Hindi, Khari Boli and Maithili), integrated prose and poetic forms to render multiple temporalities. In *Maila Anchal*, several Sanskrit words are phonetically altered to make it intelligible for the rural readers. Hansen (1981: 281) points out that, 'Renu reveals a linguistic virtuosity unrivaled in modern Hindi literature. The patchwork of diverse ethnic groups that constitute the population of Purnea District, far in the northeastern corner of Bihar bordering on Nepal and West Bengal, has been recreated in the linguistic texture of his fiction.' By bringing the oral traditions, Renu assimilates *Maila Anchal* with indigenous techniques of storytelling. The mixing of story genres, the most pronounced being the *katha*, *qissa*, *git*, and *gatha*, creates different temporalities within the novel. Indeed, these are the sedimented forms that Ricoeur argued as central to the history of narratives. Ricoeur (1992: 163–4) referencing Benjamin points out that, 'Walter Benjamin recalls that, in its most primitive form, still discernible in the epic and already in the process of extinction in the novel, the art of storytelling is the art of exchanging *experiences'*.

Srilal Shukla's *Rag Darbari* is built around a series of interconnected events in an imaginary village, Shivpalgunj, in Uttar Pradesh. The stories are organized around Vaidyaji (an ayurvedic physician, hence the name), who as a local politician and a self-professed devotee of Gandhi, embarks on spreading development and democracy in the village. Through multiple overlapping layers of narratives of village life—vignettes from the everyday chatter and gossip, the machinations in local governance institutions, the village farmers cooperative, etc.—Shukla provides socio-political commentary. Written as a comical satire of village politics, the novel is peopled with characters throughout the interlinked stories. In Rupert Snell's (1990) account, *Rag Darbari's* genre and style combine disparate linguistic elements where Hindi words are mixed with English ones. Since English is not comprehensible in village settings, 'its incomprehensibility makes it an iconic symbol rather than a means of communication; it develops a semiotic potency equivalent to that of the Sanskrit *sloka*' (Snell 1990: 164).

Indeed, as in *Maila Anchal, Rag Darbari* demonstrates the significance of the acoustic effect created by the juxtaposition of words from different languages on the ears of the people. Shukla's recourse to irony, humour, and satire, the references to Indian aesthetic and religious idioms and metaphors, imbued with realistic representations of characters, events, and happenings, enable him to develop a kind of 'comic realism'. As Ulka Anjaria (2006) notes, through the politics of 'ulti batein'—that is, the inversions of conventional realistic descriptions, presenting effects before causes, appearances before essences—imbued in the plot structure of the novel, Shukla undertakes a critique of the postcolonial state.[15] Throughout *Rag Darbari*, 'the language of refractions, bends, travesties and inversions captures the abject experience of the state...' (Anjari 2006: 4797), while noting the failed promises of the nationalism and Nehruvian state.

Ricoeur's hermeneutic insights on narrative, temporality, and social imaginaries allow us to discern how language, words, and speech in the television serials acquire a aural–visual density, wherein the identities of the characters and protagonists, tied to various social practices—via actions, performances, and events—are structured around a whole repertoire of idioms, imaginations, symbols, and 'ways of being' marked by the presence of multiple temporalities. Indeed, the social imaginaries in *Maila Anchal, Rag Darbari*, and *Godan*—in their novel form as well as the television adaptations—occupy a 'fluid middle ground between embodied practices and explicit doctrines', most notably ideologies, and the idioms of imaginaries are 'expressed and carried in images, stories, legends, and modes of address' (Gaonkar 2002: 10). To this end, the notion of social imaginaries opens up an important avenue for studying narrative identity and temporality in the novels and their television adaptations. Therefore, the notion of social imaginaries developed in Ricoeur's writings, and by the postcolonial scholarship, enables a creative and critical interpretation of how in the three novels, tradition is refigured at the level of aesthetics, and not as an ideology alone. Premchand, Renu, and Shukla, through particular linguistic strategies, decode a range of 'mythic-realist' elements. In doing so, they make narrative and human action interchangeable. Although the televisual representations of the actions and events translated from the novels lose the radical edge, there are a significant number of moments in the serials that stand out.

While the television adaptations of *Rag Darbari* and *Maila Anchal* reproduced the novels' narrative forms, maintaining fidelity to the plot structures of the literary works, the naturalist principles of the television medium define the unfolding realism in the serials.[16] This dominance of naturalism tends to dissipate the critical force of irony and humour. However, significant moments in both *Rag Darbari* and *Maila Anchal* represent a break from this sort of naturalism as their language and form and mixing of story genres disrupt it. For instance, *Rag Darbari*'s 'ulti batein' reverses the logic of realism and frames the comic moments as satire. In *Rag Darbari*, while the actions, events, and plots are organized around the characters, Vaidyaji (Manohar Singh), Ruppan Babu (Veerendra Saxena), and Ranganath (Om Puri), they retain fidelity to the novel's language of 'comic realism'. Similarly, in *Maila Anchal*, words and sentences acquire colloquial and visual density, frequently transferring meanings from one linguistic register to the other. The idiomatic grammar of Renu inflects the speech of the characters in the series, particularly the protagonist Dr Prashant and Kamili and many others. Unlike in the novel, the emotions and affective states of the characters are enacted and become visible in the televisual version as bodily gestures. Renu's strategy of grafting words and idioms from one language to another (Sanskrit, Hindi, Khari Boli, and Maithili), integration of prose and poetic forms to shape the narrative structure of the novel, enabled him to develop a sharp critique of the workings of the newly independent Indian state.

In *Maila Anchal*, Renu mixes genres like katha, qissa, git, and gatha into the novel form. He invokes pre-capitalist and mythic narrative forms like songs and ballads that pre-date and militate against a quintessentially modern genre—the novel. These storytelling modes embody a different concept of time (cyclical and heterogeneous time) as against the linear, homogeneous time of the nation. Thus, Renu is able to bring multiple temporalities into play in his work. How does one, then, explain the presence of heterogeneous time? Dipesh Chakrabarty (2000), developing his critique of Benedict Anderson's account of nation as an 'imagined community' located in 'homogeneous empty time', argued that social and religious practices in India cannot be grasped by remaining within the ambit of a linear and singular conception of time; in fact, these practices belong and emerge from an uneven and dense 'heterogeneous time'. In his philosophical mediation on narrative, Ricoeur (1984) noted that narratives embody two dimensions of temporality: a chronological, episodic dimension, where a story is made up of events; and a non-chronological, configurational dimension, where the plot organizes significant wholes out of scattered events. Whereas the episodic dimension is marked by a linear representation of time, the configurational points towards 'within-time-ness', a phenomenological notion of time that is experienced in terms of past, present, and future. The linear and phenomenological conceptions of time, Ricoeur argued, are integrated, and that it is narrative that has a capacity to explain the complex interplay of time. Against the singular conception of the 'homogeneous empty time', I argue that Chakrabarty's and Ricouer's works offer a way forward in understanding the multiple temporalities through which people imagine their collective social life.

The identities of various characters and protagonists in the televisual *Maila Anchal*, *Rag Darbari*, and *Godan* can be analytically grasped through what Ricoeur (1992) has characterized as a person's narrative identity. Ricoeur has argued that a person's narrative identity can be approached via two interconnected and overlapping notions of identity: *idem* (sameness) and *ipse* (selfhood).

While idem identity refers to 'sameness of body and character, our stability illustrated by genetic code', ipse identity pertains to our 'selfhood, the adjustable part of our identity', and furthermore, the two kinds of identities—of sameness and difference—offer coherence to the self and the possibility for change and reflexivity (Ricoeur 1992: 113–68). In the televisual *Rag Darbari*, the identities and life-worlds of the main protagonists, Vaidyaji, Ranganath, and Ruppan Babu, in *Maila Anchal*, Dr Prashant and Kamili, and in *Godan*, Hori, Dhaniya, and Gobar, as well as their wide network of relationships with peasants and rural community are layered with phenomenological detail. Their embodied identities become visible in the unfolding actions, events, and plotlines that centre around Vaidyaji's machinations in *Rag Darbari*, Hori's plight in *Godan*, and Dr Prashant and Kamili's relationship in *Maila Anchal*.

In the Indian situation, several pre-capitalist performative practices and traditions, along with the colonial, postcolonial, and capitalist discourses, have shaped the development of broadcast and televisual forms. In the televisual adaptation of the literary works, the narrative forms of the novel undergo significant changes, particularly through the formal procedures of the audiovisual medium. The televisual adaptation of *Godan*, by Gulzar, follows the main thrust of Premchand's critique, but the director prefigures and 'emplots' specific aspects of the literary work, and structures it to fit the audiovisual format of the medium. Talking about this, Gulzar noted: 'I have designed these two stories [*Godan* and *Nirmala*] as full-fledged films because the episodic structure of a television serial would have affected the continuity of the novels' (*The Tribune*, 2004). By taking creative liberties with Premchand's *Godan*, Gulzar restructures particular narrative elements from the novel to fit the televisual form of the mini-series. Indeed, the filmmaker took out a portion of the novel, focusing on the story of Hori and Dhania in the village, ignoring the city segments in which they are absent. For Gulzar, the aesthetic and formal decision to structure the novel for the televisual also has a political intent: 'the state of our villages has only worsened. Farmers still commit suicide and starving women are forced to sell their children for a pittance. We must therefore listen to Premchand's timeless voice.'

In the televisual version of *Godan*, the plight of Hori and his family is represented through a realist form shaped largely by the formal elements of the television medium. The 'radical' realism of Premchand's text is substituted by a 'progressive' realism of Gulzar's televisual rendition. However, Premchand's novel, *Godan*, and its televisual adaptation by Gulzar are shaped in relation to different and interconnected political conjunctures. Premchand's work engaged the inability of the 1930s anti-colonial nationalist politics in the face of increasing encroachment of usury forms of capital and debt financing structured by the semi-feudal and colonial relations, whereas Gulzar sought to relate this to the postcolonial situation of peasants in rural India. In fact, John Caughie's (2000:109) elaboration of 'progressive realism' in the context of British television presents some ideas that may be relevant to Indian television, particularly the development of aesthetic and cultural forms for the televisual medium:

> television's progressive function may indeed be to bring the world into the home in ways which sometimes escape the order of institutional discourse—and sometimes don't. At the same time, within the more conventional and conservative forms of representation (whether they be naturalistic, realistic, or modernistic) the new elements of social discourse may enter the stage of public visibility, as if they have always been there, adding to social discourse without transforming it or moving it on. Content which is innocent of form guarantees nothing—but neither does form alone.

Several Indian filmmakeᵢᵤ and television producers have experimented with the televisual form in adapting novels, short stories, and plays.[17] Mani Kaul's *Ahmaq* (Fyodor Dostoevsky's, The Idiot) is an instance where the naturalist, realist, and modernist elements are deployed in reworking the formal, aesthetic, and narrative form of the novel for television. James Monaco (2000), referring to the development of television forms in the American context, points out that the series as a basic unit of television enables the development of plot and character, and it was the mini-series or novel-for-television style that revived television programming as serial format. The literary-based serials discussed earlier in this chapter, and adapted from novels, short stories, and plays, reveal such interesting reworking of the narrative forms that point to an emergence of a particular style/genre in the Indian context where the pre-capitalist performative traditions overlap with the realist, melodramatic, and modernist forms.[18] A few serials point to a creative 'refiguration' of tradition combining critical and progressive force of the various aesthetic forms.

Although the preceding discussion has been preliminary for the most part, my modest attempt has been to explore possible points of convergence between Ricoeur's philosophical hermeneutics and neo-Marxist and postcolonial approaches in examining televisual narrative forms. The serials adapted from literary works constitute an important group of narratives that stand out from Doordarshan's programming 'flow' and yield several important insights in thinking about the relationship between postcoloniality and television. In the present moment of globalization and 'neoliberal' conjuncture in India, and in the larger contexts in which scholars have characterized television as part of the machinery of capitalism, it is crucial to develop critiques to *engage forms of television*.

Notes

1. Geeta Kapur (2000) interpreted myth in terms of Ricoeur's philosophical hermeneutics. Through an examination of two films, *Sant Tukaram* and *Devi*, she outlined several crucial narrative forms that emerge and are built upon myth: 'Recuperation of tradition is not just an ideological operation; that it must be perceived at the level of aesthetics proper. We should recognize above all the narrative strategy of such transformations whereby an inherited iconography is decoded and sometimes radicalized. For the narrative move is most closely analogous to, indeed interchangeable with, human action' (Kapur 2000: 233).

2. Andrew (1984) and Stadler (1990), among others, noted this lack of attention to Ricoeur's work in film and television studies. However, there have been some analyses of films that engaged aspects of Ricoeur's work (Andrew, 1993; Rosen, 1993). In her recent book, Caroline Bassett (2007) developed an interesting analytic framework that combined aspects of Ricoeur's narrative model of mimesis with Jameson's narrative approach to study how new media narratives are shaped by a range of discourses.

3. Discussing Amitav Ghosh's (1995) English novel, *The Calcutta Chromosome*, and building insights from Dipesh Chakrabarty's and

Amitav Ghosh's correspondence on the question of vernacular literary and religious traditions as crucial resources, Ghosh (2004) develops an interesting argument regarding how Amitav Ghosh 'grafted' a larger vernacular tradition of fiction writing (Rabindranath Tagore and Phanishwarnath Renu) onto his English novel. I would suggest that the televisual adaptations of literary works on Doordarshan be considered in terms of engaging vernacular literatures and idioms as well.

4. From a different perspective, the American film scholar and philosopher Stanley Cavell (1984: 85), contrasting cinema with television, indicated that 'if television's aesthetic practices are located in the serial–episode mode of composition then an investigation of the fact of television ought to contribute to understanding why there should be two principles of aesthetic composition'.

5. Tracing the beginnings of television series and serials in the naturalistic drama of Henrik Ibsen and August Strindberg, and nineteenth and twentieth centuries serialized fictions, Williams (1974: 54) noted that the longer formats proved particularly useful to television programme planners. However, this television 'form' was also shaped by other constraints for writers and producers who 'usually find themselves writing within established formation of situation and leading character, in what can be described as collective but is more often a corporate dramatic enterprise'. Williams' discussions of television relate to Britain and the United States (US) in the mid-1970s, and indeed are pertinent to the current moment of media globalization within which television 'flows' and 'forms' circulate with increasing frequency between and across North America and the rest of the world. A majority of television serials and soap operas have formulaic storylines and plots, and in many instances, writers script individual scenes in terms of 'beats' to hold the attention of audiences/viewers. The serials are infused with consumer sensibilities and are embedded as commodity forms within the larger 'flow' of programming. See Newman (2006) and Altman (1986).

6. The first televisual serial mode of composition for Doordarshan was initiated in 1976 with a serial called *Aisa Bhi Hota Hai*. I thank Abhijit Roy for bringing this to my attention. However, it was with *Hum Log* that the serial mode of composition became a regular feature of television programming. For the design and implementation of the serial-mode structure, *Hum Log* drew upon Mexican and other Latin American 'pro-development' genre of programmes as well.

7. During 1978–82, when Doordarshan began its broadcasting services on a limited basis all over India, films became a significant part of the programming. On Saturdays, 'regional' language films were transmitted

from the respective regional networks, and on Sundays, a song-and-dance episode (*Chitrahaar*), followed by a Hindi film, was transmitted on the national network. Since the 1990s, however, the film-based programming has evolved into numerous genres/shows: movie clips featuring dialogues, comedy acts, specific music forms, etc., interviews with film personalities in the form of talk shows, and a whole gamut of programmes.

8. Within television studies, realism is understood as the defining aesthetic of television based on the medium's institutional and journalistic modalities. John Corner (1992) notes that the influence of film theory also shaped the centrality of realism to television. Williams's discussion of television as a combination of various cultural forms along with the medium's own mixed forms offers important analytic possibilities in pursuing the connections between realism and television.

9. A narrative analysis ought to begin with the study of the formal properties of the story and discourse before interpreting the meanings embedded within the narratives. Narrative is broadly composed of a story and a discourse: story is comprised of *events* and *existents*, which in turn are made up of *actions, happenings, characters* and *settings*. The discourse, on the other hand, is elaborated through the *narrative transmission* and *manifestation* of the components of the story in specific ways. Thus, examination of story and discourse reveals the ideological tasks of the narrative. Television programmes are constructed through collaboration among several people—producers, directors, camera persons, set designers, script writers, editors, artists, technicians, etc. These people deploy various strategies of storytelling that is encoded at the formal and ideological levels.

10. Andrew's (2000: 44–5) essay and review of Francois Dosse's French biography on Ricoeur offer a brilliant commentary on Ricoeur's work—the formative influences, intellectual development of his thought, and the remarkable philosophical career. According to Andrew, 'to calculate the impact of his workaday ethic, it is enough to note that Ricoeur directed the thesis or served as mentor to Jean-Luc Nancy, Jacques Derrida, Jacques Rancière, Vincent Descombes, Jean-François Lyotard, and Michel de Certeau'.

11. Andrew (1984, 1993, 2000) notes that both within Marxism in general, and film theory in particular, Ricoeur's writings were ignored as well: 'film scholars who flirted with structural and psychoanalytic methodologies and yet who worried lest these approaches utterly smother the films they sought to interpret, could have used Ricoeur as a staunch ally. This became perfectly evident in 1977, with the appearance of *The Rule of Metaphor*—whose heart beat to the vibrant but unpredictable rhythm of creativity and art. Ricoeur was again timely (p. 119).'

12. Hayden White (1978) had offered a similar perspective regarding the relationship between history and narrative; however, there are significant differences between Ricoeur and White. Discussing the question of the reality of past, postcolonial scholar Prasenjit Duara (1993) demonstrated the importance of Ricoeur's discussion of how history and fiction emplot or refigure events.

13. Here is Premchand from Madan Gopal's (1964: 395) biography: 'if life were to be portrayed as it actually was, it would become only a photograph of life. For the writer, unlike the sculptor, it is not necessary to be a realist, even though he may attempt to be one. Literature is created to enable humanity to march ahead, to elevate its level of existence. Idealism has to be there, even though it should not militate against realism and naturalism. Naked realism would be more than a police report, and naked idealism no more than a speech from a pulpit. What is needed is a blend of the two.'

14. Chandra (1982: 612–13) presents a cogent discussion of the complexity of Premchand's work and its critique of nationalism: 'But Godan offers no hope; not even contact with nationalists, for whatever it is worth. Considering that Godan was begun in 1932 and continued during the kisan agitation launched by the U.P. Congress, the political insulation of Semari and Belari cannot be related to the contemporary political situation. A more likely explanation seems to be Premchand's determination—evidenced by the sustained austerity of the rural portions of the novel—to permit no false hopes this time. The lone occasion when some hope is offered is when Gobar, the rebellious son of Godan's protagonist, returns to the village from Lucknow. But the angry young man can achieve little beyond an impotent vent of collective anger in a hastily put up farce. And he, too, goes back to Lucknow. So complete, in fact, is this insulation that even when people like Prof. Mehta and Miss Malati, who are politically active and alive in the city, visit these villages, they do so for either recreation or social service.'

15. In his essay on narratives of corruption, Akhil Gupta (2005) considers Shukla's literary novel, *Rag Darbari*, an evocative and quasi-ethnographic representation of village life. Through his contextual analysis of the representations of corruption in Shukla's literary work, and F.G. Bailey's ethnographic research on Orissa's village Bisipara, Gupta demonstrates the workings of local institutions, bureaucracy, and developmental and nationalist discourse in two different genres of writings. In an interesting discussion on the close relationships between literary fiction and anthropological research on development, David Lewis *et al.* (2006) demonstrate the significance of literary representations in relation to the scholarly one's.

16. I am using the term 'naturalism' broadly and as outlined in television aesthetics and production handbooks. Williams (1977) and Caughie (2000) noted that television has a propensity towards natural-ist–realist method since the formal language and graphic design of the medium embodies the naturalistic principles as a mode of composition. For Williams and Caughie, television in the West developed by drawing upon drama and various cultural forms. Television dramas in India have not inherited the shared legacies as in Britain. Major aesthetic influences have been radio and cinema. For an interesting discussion of how radio and television shaped Urdu drama in Pakistan, see Alain Desoulieres (1999).

17. As noted earlier, during the 1980s, Doordarshan enlisted several filmmakers to produce programmes on the nationalist movement. Most of these programmes were propagandist, serving the government's inter-ests. The noted filmmaker, Kumar Shahani, who was unable to get his documentary—produced for the Madhya Pradesh government—screened on Doordarshan pointed out that 'many of our film-makers—some well-known names included—have compromised and started making "slottable" films to suit the needs of the system'. However, Shahani did acknowledge that several good programmes were transmitted for a brief period. See Shahani (2002: 13).

18. While I provide examples of television serials based on literary works in Hindi transmitted on Doordarshan's national network, there are several other serials in various Indian languages that have been transmit-ted on the regional television networks as well. Rao (1995) examined a serial based on the Marathi novel, *Paulakhuna*, transmitted on Bombay Doordarshan.

References

Altman, Rick. 1986. 'Television/Sound,' in Tania Modleski (ed.), *Studies in Entertainment: Critical Approaches to Mass Culture*. Bloomington: Indiana University Press.

Andrew, Dudley. 1984. *Concepts in Film Theory*. London, New York: Oxford University Press.

———. 1993. 'History and Timelessness in Films and Theory', in David Klemm and William Schweiker (eds), *Meanings in Texts and Actions: Questioning Paul Ricoeur*, pp. 117–26. Charlottesville and London: University Press of Virginia.

———. 2000. 'Tracing Ricoeur', *Diacritics*, 30(2): 43–69.

Anjaria, Ulka. 2006. 'Satire, Literary Realism and the Indian State: *Six Acres and a Third* and *Raag Darbari*', *Economic and Political Weekly*, 41(46): 4795–800.

Bassett, Caroline. 2007. *The Arc and the Machine: Narrative and New Media*. Manchester and New York: Manchester University Press.

Bordwell, David. 1985. *Narration in the Fiction Film*. Madison: University of Wisconsin Press.

Caughie, John. 2000. *Television Drama: Realism, Modernism, and British Culture*. Oxford: Oxford University Press.

Cavell, Stanley. 1984. 'The Fact of Television', *Daedalus*, 111(4): 79–96.

Chakrabarty, Dipesh. 2000. *Provincializing Europe: Postcolonial Thought and Historical Difference*. Princeton and Oxford: Princeton University Press.

Chandra, Sudhir. 1982. 'Premchand and Indian Nationalism', *Modern Asian Studies*, 16(4): 601–21.

Corner, John. 1992. 'Presumption as Theory: "Realism" in Television Studies', *Screen*, 33(1): 97–102.

Das, Veena. 1995. 'On Soap Opera: What Kind of an Anthropological Object is it?', in Daniel Miller (ed.), *Worlds Apart: Modernity through the Prism of the Local*, pp. 169–89. New York: Routledge.

Dawar, Jagdish Lal. 1996. 'Representations of Popular Culture in Premchand's Works', *Social Scientist*, 24(4–5): 109–29.

Desoulieres, Alain. 1999. 'A Study of Kamal Ahmad Rizvi's Urdu TV Drama *Alif Nun*', *Annual of Urdu Studies*, 14: 55–72.

Dienst, Richard. 1992. 'Image/Machine/Image: On the Use and Abuse of Marx and Metaphor in Television Theory', in Jane Gaines (ed.), *Classical Hollywood Narrative: The Paradigm Wars*, pp. 313–39. Durham and London: Duke University Press.

Duara, Prasenjit. 1993. 'Bifurcating Linear History: Nation and Histories in China and India', *Positions*, 1(3): 779–804.

Feuer, Jane. 1986. 'Narrative Form in American Network Television', in Colin MacCabe (ed.), *High Theory/ Low Culture: Analyzing Popular Television and Film*, pp. 610–19. New York: St. Martin's Press.

Gaonkar, Dilip Parameshwar. 2002. 'Toward New Imaginaries: An Introduction', *Public Culture*, 14(1): 1–19.

Ghosh, Bisnupriya. 2004. 'On Grafting the Vernacular: The Consequences of Postcolonial Spectrology', *Boundary 2*, 31(2): 197–218.

Gopal, Madan. 1964. *Munshi Premchand: A Literary Biography*. New Delhi: Asia Publishing House.

Gupta, Akhil. 2005. 'Narratives of corruption Anthropological and fictional accounts of the Indian State.' *Ethnography* 6(1): 5–34.

Hansen, Kathryn. 1981. 'Renu's Regionalism: Language and Form', *Journal of Asian Studies*, XL(2): 273–94.

Hardinge, H.R. 1934. 'Broadcasting for Rural India', *The Asiatic Review*, XXX(104): 618–23.

Jameson, Fredric. 1981. *Political Unconscious: Narrative as a Socially Symbolic Act*. New York, Ithaca: Cornell University Press.

———. 2009. 'Ricoeur's Project', in Fredric Jameson, *Valences of Dialectic*, pp. 484–532. New York: Verso.

Kapur, Geeta. 1987. 'Mythic Material in Indian Cinema', *Journal of Arts and Ideas*, 14–15: 79–108.

———. 2000. *When was Modernism? Essays on Contemporary Cultural Practice in India*. New Delhi: Tulika Press.

Kearney, Richard. 2004. *Paul Ricoeur: The Owl of Minerva*. Hampshire, England: Ashgate.

Kermabon, Jacques and Kumar Shahani. 1991. *Cinema and Television: Fifty Years of Reflection in France*. Hyderabad and New Delhi: Orient Longman.

Lewis, David, Dennis Rodgers, and Michael Woolcock. 2006. 'The Fiction of Development: Literary Representation as a Source of Authoritative Knowledge.' *The Journal of Development Studies*, 44(2): 198–216.

Longhurst, Brian. 1987. 'Realism, Naturalism and Television Soap Opera', *Theory, Culture and Society*, 4: 633–49.

Mc Arthur, Colin. 1978. *Television and History*. London: BFI.

McNay, Louis. 1999. 'Gender and Narrative Identity', *Journal of Political Ideologies*, 4(3): 315–36.

Monaco, James. 2000. *How to Read a Film: Movies, Media, Multimedia (Language, History, Theory)*. London, New York: Oxford University Press.

Newman, Michael Z. 2006. 'From Beats to Arcs: Toward a Poetics of Television Narrative', *The Velvet Light Trap*, 58(Fall): 16–28.

Prasad, Madhava. 1997. 'Television and National Culture', *Journal of Arts and Ideas*, 32–3: 120–9.

Premchand, Munshi. 1936 [1982]. *Godān*. Delhi: Rupa & Company.

Rao, Anupama. 1995. 'Televisions, Maharastrian Social Reform and Literary Imagination: Bombay Doordarshan's *Paulakhuna*', *Economic and Political Weekly*, 30(10): 521–6.

Renu, Phanishwarnath. 1954. *Mailā Anchal*. Delhi: Rajkamal Prakashan.

Ricoeur, Paul. 1968. 'Structure, Word, Event', *Philosophy Today*, 12(2): 114–29.

———. 1979. 'Ideology and Utopia as Cultural Imagination', in Donald Borchert and David Stewart (eds), *Being Human in a Technological Age*, pp. 115–28. Athens, Ohio: Ohio University Press.

Ricoeur, Paul. 1984. *Time and Narrative*, Vol. 1. Chicago and London: University of Chicago Press.

——. 1986. *Time and Narrative*, Vol. 2. Chicago and London: University of Chicago Press.

——. 1988. *Time and Narrative*, Vol. 3. Chicago and London: University of Chicago Press.

—— (trans. by Kathleen Blamey). 1992. *Oneself as Another*. Chicago and London: University of Chicago Press.

Rosen, Philip. 1993. 'Traces of the Past: From Historicity Film', in David Klemm and William Schweiker (eds), *Meanings in Texts and Actions: Questioning Paul Ricoeur*, pp. 67–89. Charlottesville and London: University Press of Virginia.

Roy, Abhijit. 2005. 'The Apparatus and its Constituencies: On India's Encounters with Television', *Journal of the Moving Image*, 4: 1–31.

Roy, Abhijit. 2008. 'Bringing up TV: Popular culture and the developmental modern in India.' *South Asian Popular Culture*. 6(1): 29–43.

Shahani, Kumar. 2002. 'Musings of a Marxist', *The Hindu*, 12 May.

Shukla, Srilal. 1968. *Rāg Darbārī*. New Delhi: Rajkamal Prakashan.

Snell, Rupert. 1990. 'Rural Travesties: Shrilal Shukla's *Rāg Darbārī*', *The Journal of Commonwealth Literature*, XXV(1): 156–79.

Sprinker, Michael. 1989. 'Marxism and Nationalism: Ideology and Class Struggle in Premchand's Godan', *Social Text*, 23: 59–82.

Stadler, Harald. 1990. 'Film as Experience: Phenomenological Concepts in Cinema and Television Studies', *Quarterly Review of Film and Video*, 12(3): 37–50.

The Tribune. 2004. 'Gulzar's Vision of Timeless Classics', Interview with Saibal Chatterjee, 15 August.

Thompson, John B. (ed.) 1981. *Paul Ricoeur: Hermeneutics and the Human Sciences*. New York: Cambridge University Press.

Valdes, Mario J. (ed.). 1991. *A Ricouer Reader: Reflection and Imagination*. Toronto and Buffalo: University of Toronto Press.

White, Hayden. 1978. *Tropics of Discourse: Essays in Cultural Criticism*. Baltimore: The Johns Hopkins University Press.

Williams, Raymond. 1974. *Television: Technology and Cultural Form*. Hanover and London: Wesleyan University Press.

——. 1977. 'A Lecture on Realism', *Screen*, 18(1): 61–74.

4

Spaces of Television

Rethinking the Public/Private Divide in Postcolonial India

SHANTI KUMAR

In the growing literature on media globalization in India—and elsewhere around the world—the debate often focuses on whether private forces in the marketplace should be allowed to dominate what is essentially the public domain of television (see, for instance, Banerjee and Seneviratne 2006 and Price and Verhulst 1998). According to some scholars, the public service mission of Indian television is in crisis due to the upheavals caused by globalization and media privatization. In their 2006 book, *Public Service Broadcasting in the Age of Globalization*, Indrajit Banerjee and Kalinga Seneviratne are critical of the rise of infotainment in television programming which, they argue, is the result of excessive commercialization and the growing corporate control of media. In *News as Entertainment: The Rise of Global Infotainment*, Daya Kishan Thussu (2007) is critical of what he describes as the Bollywoodization of television news as a result of the growing commercialization of television in India. Arguing that three

Cs—cinema, crime, and cricket—define most of the content in television news, Thussu (2007) is concerned that the ideological imperatives of infotainment in commercial television channels are debasing the quality of public deliberations in Indian democracy.

John Sinclair (2005), on the other hand, argues that one of the most unexpected consequences of globalization and media privatization has been the 'Indianization of Indian television', thanks to the rise of regional-language channels such as Asianet in Malayalam, Eenadu TV in Telugu, and Sun TV in Tamil. The profusion of private television channels in a variety of Indian languages, Sinclair argues, has not only suffused Indian television with linguistic and regional diversity to disrupt the hegemony of Doordarshan's ideological agenda, but has also been a stimulus for Bollywood and other regional-language film industries. Sinclair (2005) finds that the convergence of the Indian television and film industries has led to the creation of new programming genres—such as reality shows based on song-and-dance sequences from films—and hybrid formats that have Indianized conventional television genres such as sitcoms, dramas, and soap operas.

The dramatically divergent conclusions drawn by media scholars like Thussu and Sinclair in their analysis of globalization and media privatization in India are significant because they point to the great variation and confusion in definitions of the public and the private in India. In this chapter, I critically evaluate the many definitions of the 'public' in Indian television in relation to discourses of globalization and media privatization. I also discuss how the nationalist agenda of broadcasting defined the 'public' in Indian television from the early 1950s to the late 1980s. Moreover, I examine how the demands for alternative models of broadcasting, and the rise of private commercial satellite channels since the 1990s, have disrupted Doordarshan's hegemony over definitions of the 'public' in Indian television.

In this chapter, I draw on feminist and postcolonial theories to argue that the private is not the binary opposite of the public, but is, in fact, crucial to and constitutive of what we define as the public sphere. I discuss how representations of traditionally private desires of sexuality and intimacy in soap operas, reality TV shows, and music television are redefining the public in India. I also outline the ways in which private desire is made visible—and thus made public—through the convergence of the television screen, the cinematic screen, the computer screen, and the mobile screen. In the final part of the chapter, I deconstruct the Eurocentric biases of the central tenets of public sphere theory to show how category systems such as 'public' versus 'private' are highly problematic in the study of television in global contexts. I conclude by demonstrating how discussions of 'public' versus 'private' television force us into an either/or debate, even though such category systems are always–already hybrid in the culturally heterogeneous and ideologically overdetermined terrain of globalization in postcolonial societies such as India.

Interrogating the Public/Private Divide

In 'Public/Private: The Limitations of a Grand Dichotomy', Jeff Weintraub (1997b) outlines four 'models' that have conventionally been used to distinguish public from the private: (a) a liberal–economic model that defines the public as state administration and the private as the market economy; (b) a 'civic model' in which the public means community and citizenship, as distinct both from state sovereignty and market economy; (c) a 'sociability' model, which sees the public sphere as a fluid realm of symbolic display and theatrical self-representation and is distinct from both formal structures of social organization and private domains of informality, intimacy, and domesticity; and (d) a feminist model which critiques the traditional opposition between the private defined as the domestic or the familial and the public defined in terms of the dominant political–economic order of state interests and market forces.

The 'grand dichotomy', as Weintraub (1997) puts it, has confounded many philosophers, political scientists, policymakers, social activists, and citizens who have struggled to clearly delineate the boundary between the public and the private spheres in society. As Bruce Robbins (1993) points out, this struggle is as much evident in Walter Lippmann's unhappy search for the 'sovereign and omnipotent citizen' of the modern free world, as in Jurgen Habermas's lament for the lost ideals of the bourgeois public sphere of early modern Europe; as unmistakable in Alan Bloom's nostalgia for a postmodern Socrates and an Athenian agora in American educational institutions, as in John Dewey's romance with the lore of feudal communities of yore. Using Weintraub's aforementioned classification, Robbins foregrounds the at-once 'potent and profitable' appeal of the ideal of a universal public sphere (clearly opposed to a private sphere) that sustains the rhetoric and politics of liberal democracy. He writes: 'the rhetoric that seeks to inflame its hearers against a perceived loss or dimunion of the public is so potent and profitable ... [because] [t]his rhetoric creates an illusion of unity among people who, it now appears, rightly belong on opposite sides' (Robbins 1993: xiii).

Those like Walter Lippmann, who acknowledge the limitations of the public/private divide, find themselves, as Robbins points out, in a frustrating phantasmagoria inhabited by a phantom public. On the other hand, those who are seduced by the ideals of a universal public sphere uncritically embrace the arbitrariness of the 'grand dichotomy', and thus end up with a 'mythic town-square in the sky' that has 'nothing to do with the realities and limits' of the societies they romance or 'with the possibilities to extend our own' (Robbins 1993: ix). All they succeed in, Robbins argues, is to resurrect an awe- and fear-inspiring 'Spirit of the Past', and fail to ask crucial questions like: for *whom* was the public sphere once more public than now? Furthermore, Robbins asks: 'Was it ever open to the scrutiny and participation, let alone under the control, of the majority? Was there ever a time when intellectuals were really authorized to speak to the people as a whole? If so, where were the workers, the women, the lesbians, the gay men, the African Americans?' (Robbins 1993: viii).

As Robbins points out, many feminist scholars, revisionist historians, poststructuralist, postcolonial critics have been critical of 'the lament for a lone lost public', and have called for the deconstruction of the utopian ideals underlying the desire for a universal public sphere. With a 'qualified but steadfast commitment to the concept of the public', 'these critics move away from the universalizing ideal of a single public' and attend instead to 'the actual multiplicity of distinct and overlapping public discourses, public spheres, and scenes of evaluation that already exist, but the usual idealizations have screened from view' (Robbins 1993: xii). These critics attempt to 'detect and evaluate publicness that is already there in diverse forms' (Robbins 1993: xii). However, they also advocate caution in the project of reconstituting the public sphere in terms of the multitudinous presence of publicness in everyday life.

The cautious tone is aptly summed up by Robbins who, citing Jacques Derrida, writes, 'the fact that "public opinion" is a specter, "present as such in none of the spaces" it is held to be, does not mean one can "*simply* plead for plurality, dispersion, or fractioning"' (Robbins 1993: xii). Here, Derrida and Robbins strike at the heart of the paradox of the 'grand dichotomy' that has and continues to confound the critics and advocates of the public sphere. For, as Derrida points out, '[i]n the absence of a general forum' in a fractured public sphere, 'certain socio-economic forces may take advantage of these marginalizations' (Derrida, cited in Robbins 1993: xii). Therefore, Derrida asks, '[h]ow then to open the avenue of great debates, accessible to the majority, while yet enriching the multiplicity and the quality of public discourses, of evaluating agencies, of "scenes" or places of visibility?' (Derrida, cited in Robbins 1993: xii). The question that Derrida poses for the project of deconstructing the public/private divide is, as Robbins puts it, an 'impossible but necessary' one. It is also a question that serves as a crucial starting point in my own analysis of globalization and media privatization in Indian television.

In India, when Doordarshan was the only available network from the late 1950s to the late 1980s, the state-sponsored agenda of national broadcasting defined the many meanings of 'public' in television. During this period, the concept of 'national programming' served as the guiding framework for public broadcasting in India. For example, nationalist programmes on Doordarshan— such as *Hum Log*, *Buniyaad*, *Ramayan*, and *Mahabharat*, among others, during the 1980s—contributed to the ideological conflation of diverse meanings of publicness with state-sponsored definitions of national community and ideal citizenship (Mankekar 1999; Rajagopal 2001; Singhal and Rogers 1990).

Beginning in the early 1990s, Doordarshan's hegemony over definitions of the 'public' in Indian television was dramatically disrupted by the rise of commercial satellite channels, first, in English and Hindi and later, in several other Indian languages. Unlike Doordarshan, whose mandate as a state-sponsored broadcaster has been to serve as an agent of social change by focusing on issues like national integration, agricultural development, literacy, education, health, and family welfare, the programming strategies of private commercial channels have been driven by profit motives and the quest for larger audience share in an increasingly competitive marketplace. But since the arrival of satellite networks in the 1990s, Doordarshan has also embraced a profit-driven agenda in order to compete with the commercial channels, while some private channels have tried to carve out a niche for themselves by catering to marginalized audience groups and thus contributing to public debates on progressive social change and national development (Gupta 1998; Ninan 1995).

As is evident even from this cursory overview of television in India, state interests often do not coincide with the public interest, and commercial interests can and do sometimes serve the public interest—even though it may be an indirect outcome of the capitalist desire to gain more audiences and make greater profits. The contrast between the commercial and the public on the one hand, and the conflation of the state and the public on the other, may work in neat economic models and in liberal theories of representational politics, but they rarely work in practice due to political pressures, ideological manipulations, bureaucratic mismanagement, and so on. Therefore, to construct a binary opposition between the state-sponsored network as 'public' television and the commercial networks as 'private' channels is rather problematic in debates about globalization and media privatization in India.

This view was also embraced by the Supreme Court of India. In its landmark judgement, the Supreme Court held that the airwaves are public property that must be used in ways that ensure the expression of a plurality of views and diversity of opinions in the national community. The Court also ruled that the Government of India had a responsibility to use the airwaves to advance the citizens' rights to free speech guaranteed by the Constitution.

In its ruling, the Court explained, 'The broadcasting media should be under the control of the public as distinct from government. This is the command implicit in Article 19 (1) (a). It should be operated by a public statutory corporation or corporations' (cited in Ninan 1998: 13). On the question of broadcasting by private individuals and commercial networks, the Court ruled:

> The question whether to permit private broadcasting or not is a matter of policy for the Parliament to decide. If it decides to permit it, it's for the Parliament to decide, subject to what conditions and restrictions should it be permitted. Private broadcasting, even if allowed, should not be left to the market forces in the interest of ensuing that a wide variety of voices enjoy access to it. (cited in Chenoy 1997)

In defining the airwaves as a public property that is free from both state control and commercial forces, the judges were advancing a 'civic model' of public broadcasting where the public means community and citizenship, as distinct both from state sovereignty and market economy (Weintraub 1997b). In the institutional context of Indian television, the civic model has been systematically marginalized both by Doordarshan's centralized command-and-control structure and by the advertising-driven, market-oriented approaches of commercial channels. The Prasar Bharati Corporation, created by the Government of India, was to provide Doordarshan with some autonomy, but that hardly qualifies for a civic model of public broadcasting that is distinct from both state and market control in the Habermasian sense.[1]

There are small but growing numbers of organizations—nongovernmental organizations (NGOs), youth organizations, and religious bodies—working to create a civic model for public broadcasting in India that is distinct from both state control and market forces. Vinod Pavarala and Kanchan Malik (2007) highlight some of the work being done in the realm of community radio in India by NGOs like the Deccan Development Society (DDS) in Zaheerabad, Andhra Pradesh; VOICES/MYRADA in Budhikote, Karnataka; and the Kutch Mahila Vikas Sangathan (KMVS) in Gujarat. As Pavarala and Malik (2007) point out, each of these private NGOs uses a different model of 'public' broadcasting to empower the people of the communities it serves. The DDS, which set up a community radio station in 1998 with assistance from United Nations Educational, Scientific and Cultural Organization (UNESCO), trains poor Dalit women to produce shows, manage the local station, and disseminate audio tapes in village *sanghams* or communities. VOICES/MYRADA, which started an audio production centre, *Namma Dhwani* (our voice), in 2001—also with the assistance of UNESCO—gives rural men and women basic training to create audio programmes and transmits these programmes through a cable television link. The KMVS, which has a long history of doing development projects in the Kutch region, started airing Kutchi language programming in 30-minute slots by purchasing commercial time on All India Radio (AIR) Bhuj in 1999 (Pavarala and Malik 2007).

The DDS, VOICES/MYRADA, and the KMVS are good examples of NGOs that are neither public nor private in the liberal–economic dichotomy of state versus corporate, but are at once public and private in a civic sense. The civic model of private groups creating public or community media thus forces us to rethink the public/private distinction in India. For, what the civic model of community-based media reminds us is that the equation of the private with the corporate is highly problematic because there is much in the private sphere that is not under the direct control of corporations.

But the civic model of community-based radio is not the only way to think about non-statist and non-corporate media. As Weintraub (1997) reminds us, there is a whole world of 'public life' that is not statist or corporate, but is also not about issues of citizenship or community organizations in the civic sense. Or as Weintraub (1997: 21) puts it, 'the key to this alternative version of the "public" realm is not solidarity or obligation, but "sociability"'. The 'public life' celebrated through 'sociability' is a fluid realm 'of multi-stranded liveliness and spontaneity' that arises from 'the ongoing intercourse of heterogeneous individuals and groups that can maintain a civilized coexistence' (Weintraub 1997b: 21–2).

In the Indian context, the realm of 'sociability' in public life refers to spaces of everyday interactions such as the street corner, the neighbourhood, the *nukkad*, or the *mohalla*, or the *adda* in

Dipesh Chakrabarty's (2000a) famous formulation. These spaces of sociability—or sociality as Chakrabarty puts it—are neither public nor private in the liberal–economic sense of state versus market forces, or in the civic sense of communitarian responsibilities and citizenship, but they constitute the heart of public life in colonial and postcolonial India.

While folk media and mass media like radio, television, and film have always played an important role in public spaces of sociability, social networking sites on the Internet such as facebook, myspace and orkut, websites such as bigadda.com and nukkad.com, and phone-based short messaging services (SMS) are making it possible for people to be part of public life in new ways in cities and towns across India. Increasingly mobile phones, internet cafes, and social networking sites are transforming the ways in which people interact with each other in public spaces by going online at home or in a cyber cafe, or by talking or texting on the cell phone while sitting in a park or a while walking down the street. This transformation has induced media scholars and practitioners to rethink the mediated character of public spaces through the use of creative terms such as 'cyberpublics' (Sundaram 2000), 'portable publics' (Ghosh 2007), and 'mobile publics' (Punathambekar 2011).

Ravi Sundaram (2000) uses the term 'cyberpublics' to describe the way in which cyberspace has become an integral element of the changing nationalist imaginary in India. Arguing that cyberspace poses a significant challenge to the state-centred model of nationalism in India, Sundaram (2000) maps the emergence of three distinct but overlapping cyberpublics: (a) the state cyberpublic which seeks to reassert its hegemony in the new media environment; (b) a cyberpublic of the transnational elites who see regimes of global consumption as solutions to problems of underdevelopment; and (c) a cyberpublic of activists and members of new social movements such as the ecological movement, women's groups, and labour activists who are beginning to use cyberspace in creative ways to counter their marginalization by the dominant order.

Shohini Ghosh (2007) uses the term 'portable publics' to describe new public spaces which are emerging—albeit fleetingly—to support and disseminate the works of independent documentary filmmakers who do not have easy access to the state-sponsored network, Doordarshan, or to any of the private television channels, or to many of the commercially run film theaters in India. Aswin Punathambekar (2011) uses the term 'mobile publics' to foreground how mobile media such as public call offices, cell phones, and SMS texting are enabling fans of reality television shows like *Indian Idol* to form new types of publics as they mobilize to get votes for their favourite contestant.

What I have summarized here is a small segment of a growing body of work in Indian media studies that draws from urban studies, cultural geography, and postcolonial theory to examine the everyday spaces of sociability as the stuff of public life, but much more needs to be done in this arena. At the same time, it is important to recognize that there is also a whole world of 'private life'—intimacy, sexuality, family, and friendship—which, as Weintraub (1997) suggests, has little to with the state, corporate interests, civic concerns, or even with public spaces of sociability, but is central to our understanding of publicness.

Private Desires and Public Cultures

Many social historians, cultural critics, and feminist scholars have for long argued that the private sphere is not the binary opposite of the public sphere but is, in fact, crucial to and constitutive of what we define as the public. Postcolonial critics such as Chakrabarty (2000b), Partha Chatterjee (1993), and Lata Mani (1998), among others, have shown how the public/private distinction in India evolved very differently than in Western Europe because of colonialism. In colonial India, the public sphere was the realm of colonial administration and thus, the private realms of the home, the family, and other domestic relations became the sites where nationalists, religious leaders, patriarchal authorities, and cultural reformers struggled to define what Indian history, culture, and tradition were. In the Indian context, then, the private sphere has not been just the space of intimacy, sexuality, family, and friendship, but also the place where more traditionally public concerns of community, nationality, and solidarity have been debated and resolved. However, in the postcolonial context,

I argue that colonial distinctions of the inside and the outside, and Eurocentric notions of the public and the private, have been blurred. Once the domain of colonial authority, the public sphere is now an important site where private notions of home, family, and other domestic relations are mobilized as nationalist concerns by the postcolonial elites.

As Mankekar (2004) points out, during the 1990s, there was an eruption in representations of erotic desire on Indian television. While some scholars and media critics attribute the proliferation of erotic desire in Indian public culture to the rise of a capitalist consumer culture in the post-liberalization economy in the 1990s, others attribute it to the arrival of transnational satellite television networks like STAR TV and MTV, and the subsequent rise of home-grown soap operas, films, and talk shows produced specifically for viewers in India and its diasporas. Analysing the discourses of desire surrounding commodities, and television programmes shown on television, Mankekar seeks to understand the place of erotics in the reconfiguration of gender, family, class, and nation in the Indian public culture. While Mankekar is careful to recognize that there is a complex genealogy of the erotic in colonial and postcolonial India, she argues that there is something qualitatively and quantitatively different about recent representations of the erotic in public culture. She writes:

> recent representations of the erotic are deeply imbricated with the feverish commodity consumption precipitated by the expansion of mass culture, the liberalization of the Indian economy, and the introduction of globalized capital. Second, the production, circulation, and consumption of these representations occur in a transnational, intertextual field. Furthermore, representations of the erotic in postcolonial India frequently provoke discourses of the defense of Indian or national culture: notwithstanding older traditions of erotics, contemporary representations are often associated with Westernization and are therefore deemed transgressive. Nowhere is this association stronger than with reference to women's erotic desire. (Mankekar 2004: 408)

Mankekar (2004: 418–19) argues that Indian television shows since the 1990s have exhibited an 'unprecedented fascination with intimate relationships—particularly marital, pre-marital, and extramarital relationships—and contained new and varied representations of erotics (explicit as well as implicit)'. The programmes she refers to include soap operas, sitcoms, talk shows, made-for-television films and mini-series, music shows based on Indian films, and MTV-type music videos. Mankekar (2004: 419) finds that 'the emphasis on the intimate and the erotic was strongest in talk shows (which proliferated after the advent of transnational television), soap operas, MTV-influenced music videos, and television advertisements.'

In a media-saturated society like India, private desires have always been a public concern thanks to the popularity of commercial cinema in languages like Hindi, Tamil, Telugu, Kannada, and so on. But now, as Mankekar points out, the private has become an everyday concern of the public on talk shows, soap operas, music videos, and reality shows, where traditionally domestic issues are made visible both on and off screen.[2] While soap operas, reality TV shows, and music television are redefining the public/private divide in India, I also want to highlight the various ways in which the private realm of desire is made visible—and thus made public—through the technological convergence of the television screen with the cinematic screen, the computer screen, and the mobile screen.

In February 2010, UTV Bindass launched a youth-oriented, multimedia campaign in India called, 'What I am'. The goal of the campaign was to counter stereotypes of young people as immature and irresponsible, and to promote an image of youth culture as hip, cool, and responsible. The campaign features several young, urban Indians looking directly at the viewer/reader and proclaiming, 'Just because I'm Bindass Doesn't Mean I Do Drugs' or 'Just Because I am Bindass Doesn't Mean I Don't Believe in God', followed by the tag line, 'UTV Bindass. What I am'. *Bindaas* or *bindass* is a colloquial Hindi word meaning cool and carefree without restraint.[3]

UTV, one of India's leading media and entertainment companies since the early 1990s, launched Bindass TV in September 2007, followed by the launch of its integrated web portal, bindass.com, in July 2009. As described in the 'About Us' section of the channel's website, Bindass is 'a celebration of being young in India'.

Waxing eloquence on what it calls 'Brand Bindass!', the website goes on to claim that 'Bindass is about being Fun, Frank, Fearless and valuing Freedom in all its forms'.[4]

Bindass is, of course, not the first youth entertainment channel in India to make such exaggerated claims about its brand identity for the sake of self-promotion. MTV's slogans —such as 'I Want My MTV'—and the famous tagline on Channel [V]—'[V] are like this only'—along with similar media campaigns on many other wannabe youth channels have inundated the satellite and cable lineup in Indian television for almost two decades now with upbeat messages about their ability to represent the desires of youth cultures for freedom, rebellion, and revolution. What sets Bindass apart, according to its promoters, is that 'it is India's first 360 degree entertainment venture across television channels, mobile channels, web, gaming, merchandising, retail & nation wide ground events'.[5] Claiming that it is not just another TV channel, the website promotes Bindass as 'a platform for like minded people to come together on television, the web, mobile and at cafes'. Although its target audiences are in the commercially lucrative demographic category of youth (15–34 years), the channel also seeks to attract 'the young at heart', proclaiming that 'Bindass is all about the attitude'.[6] In redefining the lucrative demographics of the television industry from age (15–34 years) to desire ('young at heart'), Bindass seeks to recast the traditional category of audiences as marketable consumers to target audiences as attitudes and desires ('young at heart', 'bindass', 'What I am', etc.).

The globalization of traditionally national television industries and cultures, along with the digital convergence of broadcasting, cable, satellites, cell phones, and the Internet, has radically transformed the flow of television within and across cultures in recent decades. In television studies, flow is a multi-accentuated term. Traditionally, in television studies—and in the television industry—flow refers to the programming and scheduling strategies used by network executives to overcome the gaps created by commercials and publicity announcements and capture audience attention from one programme to the next, and from one segment of a programme to the next. According to Raymond Williams (1974), however, flow is less about the strategies of audience capture used by network executives and more about the ways in which audiences are able to watch television as a seamless narrative in spite of the interruptions across programmes and within programme segments. These interruptions, Williams argues, are not to be understood as mere gaps in the schedule, but as planned breaks around which discrete programmes are structured into a linear sequence that constitutes the television flow. For Raymond Williams, the phenomenon of a 'planned flow' is thus a defining characteristic of broadcasting as a technology and a cultural form that emerges from the interaction between audiences and television programming in any given viewing context.

In globalization studies too, flow has multiple definitions. In international communications, for instance, flow is defined in terms of the uneven exchanges of communication across nations due to inequities of power in global affairs. This model of flow in international communications emerges from mass communication studies, where flow is understood in terms a linear process of transmission from a given source to any receiver. Here, I am referring to source–message–channel–receiver (S–M–C–R) models, and their variations in two-step flow and multi-step flow models that have been very influential in theorizations of the politics, economics, and technologies of communication in the social sciences.[7]

More recently, in globalization studies, flow has been theorized by Arjun Appadurai (1996) as an elementary framework for understanding the cultural dimensions of globalization. For Appadurai, flow does not refer to the linear transmission of communication from a powerful sender to a relatively powerless receiver in international affairs, but to a complex, overlapping order of disjuncture and difference in the global cultural economy. To describe this new emerging order of globalization, Appadurai maps five dimensions of flow consisting of ethnoscapes, mediascapes, technoscapes, financescapes, and ideoscapes. 'Ethnoscapes' refer to movement of people as workers, tourists, students, immigrants, refugees, and others. 'Technoscapes' refer to technologies that move at high speeds across traditionally impervious boundaries. 'Financescapes' refer to rapid movements of capital on a global scale. 'Mediascapes' refer to both the global media that enable electronic transmission of information and to the variety of images that are available to audiences as resources for cultural imagination. 'Ideoscapes' are also 'concatenations of images' but are defined more explicitly as political.

Appadurai uses the suffix 'scape' to describe how the world can appear rather stable like a landscape when seen from a particular perspective, in spite of disjuncture and difference within and across the various flows of globalization. At the same time, the suffix 'scape' also allows Appadurai to point to the irregular shapes and deterritorialized movements of global flows as they work differently within and across different parts of the world. Appadurai's theorization of the cultural experience of global flows in terms of disjuncture and difference within and across the various 'scapes' bears a striking resemblance to Williams's theory of the cultural experience of television flow in terms of the commercial breaks and interruptions within and across discrete programming units in broadcasting. However, unlike Williams who considers flow as a 'planned' phenomenon of broadcasting, Appadurai considers flow as a deterritorialized phenomenon of globalization.

Much has been written about how audiences are experiencing an increasingly deterritorialized televisual culture by imagining the world as a stable landscape build around a dynamic set of disjunctive but overlapping global flows that Appadurai has theorized in terms of mediascapes, technoscapes, ethnoscapes, ideoscapes, and financescapes. But little attention has been paid to the ways in which network executives around the world are working to reterritorialize the disjunctive flows of globalization, particularly since television flow is still a planned phenomenon as described by Williams. Here, I am referring to new programming and scheduling strategies like simulcasting, multicasting, and webcasting being used by global networks to provide audiences with a seamless experience of television not only in relation to commercial interruptions but also in relation to the overlapping and disjunctive flows of globalization.

For instance, when a major American broadcasting network like ABC, NBC, or CBS simulcasts English programming in Spanish, it is a strategic attempt to reterritorialize the global flows of migrant and immigrant ethnoscapes and bilingual mediascapes into the planned flow of television in the United States (US).

Similarly, when major state-sponsored networks like CCTV in China or Doordarshan in India expand their services to reach diasporic audiences on satellite and cable channels around the world, it is a clear recognition of the growing influence of transnational ethnoscapes and technoscapes in the globalization of their national cultures. When a new media network like Bindass TV claims to provide a '360 degrees experience' by seamlessly migrating from television to a digitally convergent platform of television + cinema + Internet + cell phones + gaming +, it is yet another example of the reterritorializing strategies used by media networks to incorporate disjunctive global flows of youth cultures and their nomadic desires 'for being Fun, Frank, Fearless and valuing Freedom in all its forms'[8] into the planned flow of television.

* * *

Many of the recent changes in relations of time, space, and identity in media culture have been enormously influenced by the rapid transformations of global capitalism and technology. But as Doreen Massey (1992) points out, that is not all. Massey cautions that to reduce these changes to 'cultural logic of capitalism' (following Fredric Jameson's famous formulation; see Jameson 1990), or the logic of 'flexible accumulation' (in David Harvey's [1991] equally popular account), is to 'severely reduce their meaning and their variety' (Massey 1992: 9). Critical of influential theorists like Jameson and Harvey who uncritically accept the Eurocentric biases of the public/private distinction and use universalizing notions of time, space, place, and identity, Massey wonders if it isn't a predominantly masculine/white/First World take on things global. Massey acknowledges that there is a growing recognition in recent scholarship of the peripheral 'other'—the woman, the colonized, the marginalized, etc.—yet, she argues that there is no understanding of how from the point of view of these peripheries, the experience of time, space, place, and identity can be radically different. Massey concludes that most efforts to elaborate the relationship between time, space, place, and identity have been attempts to fix the meanings of places, to enclose and defend them; to construct singular, fixed static identities for places; to 'interpret places as bounded enclosed spaces defined through counterposition against the Other who is outside' (Massey 1992: 12).

Massey's (1992: 12) concern is that 'these economic interpretations come far too close to depriving the cultural (or the non-economic more generally) of any autonomy at all'. For Massey, some of the other strata along which the relations of time–space–identity can be articulated are ethnicity and gender. To the two important strata of ethnicity and gender that Massey identifies in her critique of Eurocentric conceptions of time, space, place, and identity, I would add another crucial stratum that many Marxist and non-Marxist critiques of the public/private divide tend to ignore: desire. What role does desire play in signifying and subjectivizing 'our' conceptions of time, space, place, and identity. Or, are 'we' to assume that time, space, place, and identity are objective, universal constants that are not influenced by desire? I would argue that in order to understand the hybrid character of the media culture in India today, one has to address the public/private distinction at the site where its paradoxical rhetoric simultaneously emanates from and disseminates to—the desiring subject. Hence, following Thomas Keenan (1993), I would argue that the 'public sphere' cannot simply be a street or a townsquare; instead, the public is in the desiring subject. Or, as Keenan (1993: 133) puts it: 'If it is any where, the public is "in" me, but it is all that is not me in me, not reducible to or containable within "me," all that tears me from myself, opens me to the ways I differ from myself and exposes me to that alterity in others'.

This redefinition of the 'public' in terms of the interiority and alterity that is both within and beyond the 'private' desiring subject is not an attempt to debunk left-wing theories of ideology critique in the public sphere and join the right-wing celebrations of individual free agency in the age of global capitalism and American-style democracy. Instead, it is an attempt to rearticulate the 'public' in terms of a multiplicity of at-once public/private spaces where desiring subjects must—and do—creatively engage with the impossible but necessary question of imagining the public sphere through a multitude of minority voices. What is at play in this rearticulation of the public is a new media assemblage of desiring subjects who creatively subvert the arbitrary distinctions of public/private, global/local, self/other, and yet are paradoxically enough always–already created by such grand dichotomies.

The 'identity' of the desiring subjects who are engendered by (and are engendering) the new media assemblage of the public/private may seem rather 'schizophrenic' to those of us more attuned to 'normal' models of identity. However, as Gilles Deleuze and Felix Guattari (1983) demonstrate, the 'nomadic' notion of multiple identities, or *haecceities*, is a much more realistic mode of individuation than the hegemonic model of hyper-normalized identity enforced in the public sphere through universalizing triangulations such as daddy/mommy/me (the familial), noblesse/bourgeoisie/plebeians (the social), thesis/antithesis/synthesis (the ontotheological), and so on.

The nomadic 'haecceity' that Deleuze and Guattari speak of is, of course, a form of individuation very different from conventional models which define humans as 'beings'. Instead, for Deleuze and Guattari, the issue is one of 'becomings'. To distil a succinct definition of haecceity from Deleuze and Guattari's writings is perhaps an exercise in futility. Instead, I will cite a short passage from their extensive writings on individuation, and yet, hopefully, not impair their many meanings. Deleuze and Guattari (1983: 287–8) write:

> There is a mode of individuation very different from that of a person, subject, thing, or substance. We reserve the name *haecceity* for it. A season, a winter, a summer, an hour, a date have a perfect individuality lacking nothing, even though this individuality is different from that of a thing or a subject. They are haecceities in the sense that they consist entirely of relations of movement and rest between molecules or particles, capacities to affect and be affected. When demonology expounds upon the diabolical art of local movements and transports of affects, it also notes the importance of rain, hail, wind, pestilential air, or air polluted by noxious particles, favorable conditions for these transports. Tales must contain haecceities that are not simple emplacements, but concrete individuations that have a status of their own and direct the metamorphosis of things and subjects. Among the types of civilizations, the Orient has many more individuations by haecceity than by subjectivity or substantiality: the haiku, for example, must include indicators as so many floating lines constituting a complex individual.

Deleuze and Guattari (1983: 289) caution us that a haecceity is not simply a dialectical synthesis of being a sedentary individual and becoming a nomad; nor is it an issue of choosing between them: 'For, you will yield nothing to haecceities unless you realize that is what you are, and that you are nothing but that'. A bit further along, Deleuze and Guattari (1983: 289–90) elaborate:

It should not be thought that a haecceity consists simply of a décor or a backdrop that situates subjects, or of appendages that hold things and people to the ground. It is the entire assemblage in its individuated aggregate that is the haecceity; it is this assemblage that is defined by a longitude and a latitude, by speeds and affects, independently of forms and subjects which belong to another plane ... At the most we may distinguish assemblage haecceities (a body considered only as longitude and latitude) and interassemblage haecceities which also mark the potentialities of becoming within each assemblage (the milieu of the intersection of longitudes and latitudes). But the two are strictly inseparable. Climate, wind, season, hour are not of another nature than the things, animals or people that populate them, follow them, awaken and sleep within them.

In formulating their theory of haecceity as a mode of individuation for nomadic desire, Deleuze and Guattari suggest that theories of 'false consciousness' perhaps blind the 'enlightened' bourgeois rationality of the critic more than the so-called plebian irrationality of the 'masses', and fables of 'Oedipal' fears haunt the adult analysts who fabricate them more than the so-called aberrant child who is subjected to these myths. By drawing on Deleuze and Guattari's theorizations of haecceity, I seek to posit that there have always been nomadic notions of public/private desires that have been written off as peripheral or deviant in the grand, totalizing, hyper-normalized narratives of Eurocentric modernity and/or postmodernity. These nomadic notions of publicness/privateness have engendered—and have been engendered by—'other' experiences of time, space, place, and identity that, at once, enable the desiring subject to assume multiple relational identities in the familial, social, and the ontotheological without suffering from a perceived sense of lack, dislocation, and placelessness.

In postcolonial India, the eruption in media representations of private desires and the digital circulation of such desires through the convergence of television with cinema, computers, cell phones, and the like has engendered a new media assemblage where colonial distinctions of the public and private are being contested and transformed.[9] What has emerged in this contested terrain is a hybrid notion of publicness which is partly defined by state-sponsored ideals of citizenship; partly dictated by market-oriented ideologies of consumption; partly determined by the technological convergence of film, television, and other electronic media; and partly driven by the diverse desires of culturally distinct audience groups in India and in the Indian diaspora. In these mediated spaces of culture and everyday life, binary category systems such as 'public' versus 'private' are highly problematic as they force us into an either/or debate even though such category systems are hybrid in the culturally heterogeneous and ideologically overdetermined terrain of globalization.

Notes

1. For an extended critique of the question of autonomy in pubic broadcasting in India, see Shanti Kumar (2005).

2. For other discussion of how soap operas, reality TV shows, and music television are redefining the public/private divide in India and in the Indian diaspora, see Sujata Moorti (2007); Divya C. McMillin (2003); Rupal Oza (2001).

3. Available at http://www.bindass.tv/.

4. Available at http://www.bindass.tv/about_us.htm.

5. Available at http://www.bindass.tv/about_us.htm.

6. Available at http://www.bindass.tv/about_us.htm.

7. For a more detailed account of the many meanings of flow in mass communications, international communications, media studies, and cultural studies, see Mimi White (2003).

8. Available at http://www.bindass.tv/about_us.htm.

9. For an analysis of the emergence of a new media assemblage in the context of globalization and digital convergence in Indian cinema, see Amit Rai (2009). While Rai focuses on Bollywood in his path-breaking book, I wish to highlight the place of television transforming notions of the public and the private in India's new media assemblage.

References

Appadurai, Arjun. 1996. *Modernity at Large: Cultural Dimensions of Globalization*. Minneapolis: University of Minnesota Press.

Banerjee, Indrajit and Kalinga Seneviratne. 2006. *Public Service Broadcasting in the Age of Globalization*. Singapore: Asian Media Information and Communication Center.

Chakrabarty, Dipesh. 2000a. 'Adda: A History of Sociality', in D. Chakrabarty, *Provincializing Europe: Postcolonial Thought and Historical Difference*, pp. 180–213. Princeton: Princeton University Press.

———. (2000b). *Provincializing Europe: Postcolonial Thought and Historical Difference*, Princeton: Princeton University Press.

Chatterjee, Partha. 1993. *The Nation and Its Fragments: Colonial and Postcolonial Histories*. Princeton: Princeton University Press.

Chenoy, Kamal Mitra. 1997. 'In Depth: The Broadcasting Bill.' The Indian Economy Overview, 1996–7, available at http://www.ieo.org/broad cast.html (last accessed on 20 May 2011).

Deleuze, Gilles and Felix Guattari. 1983. *Anti-Oedipus: Capitalism and Schizophrenia*. Minneapolis: University of Minnesota Press.

Ghosh, Shohini. 2007. 'Redefining Public Space: Ushering in the Age of New Portable Publics', *Combating Communalism*, 14(125), available at http://www.sabrang.com/cc/archive/2007/sep07/media.html (last accessed on 20 May 2011).

Gupta, Nilanjana. 1998. *Switching Channels: Ideologies of Television in India*. New Delhi and Oxford: Oxford University Press.

Harvey, David. 1991. *The Condition of Postmodernity: An Enquiry into the Origins of Cultural Change*. Malden, Massachusetts: Wiley-Blackwell.

Jameson, Fredric. 1990. *Postmodernism, or the Cultural Logic of Late Capitalism*. Durham, NC: Duke University Press.

Keenan, Thomas. 1993. 'Windows of Vulnerability', in Bruce Robbins (ed.), *The Phantom Public Sphere*, pp. 121–41. Minneapolis: University of Minnesota Press.

Kumar, Shanti. 2005. 'In Search of Autonomy: The Nationalist Imagination of Public Broadcasting', in Bernard Bel, Jan Brouwer, Biswajit Das, Vibodh Parthasarathy, and Guy Poitevin (eds), *Media and Mediations*, pp. 255–81. New Delhi: Sage Publications.

Mani, Lata. 1998. *Contentious Traditions: The Debate on Sati in Colonial India*. Berkeley, CA: University of California Press.

Mankekar, Purnima. 1999. *Screening Culture, Viewing Politics: An Ethnography of Television, Womanhood, and Nation in Postcolonial India*. Durham, NC: Duke University Press.

Mankekar, Purnima. 2004. 'Dangerous Desires: Television and Erotics in Late Twentieth-century India', *Journal of Asian Studies*, 63(2): 403–31.

Massey, Doreen. 1992. 'A Place Called Home', *New Formations*, 17 (Summer): 3–15.

McMillin, Divya C. 2003. 'Marriages are Made on Television: Globalization and National Identity in India', in Lisa Parks and Shanti Kumar (eds), *Planet TV: A Global Television Reader*, pp. 341–59. New York: New York University Press.

Moorti, Sujata. 2007. 'Imaginary Homes and Transplanted Traditions: The Transnational Optic and the Production of Tradition in Indian Television', *Journal of Creative Communications*, 2(1–2): 1–21.

Ninan, Sevanti. 1995. *Through the Magic Window*. New Delhi: Penguin Books.

―――. 1998. 'History of Indian Broadcasting Reform', in M.E. Price and S.G. Verhulst (eds), *Broadcasting Reform in India: Media Law from a Global Perspective*, pp. 1–21. New Delhi: Oxford University Press.

Oza, Rupal. 2001. 'Showcasing India: Gender, Geography and Globalization', *Signs*, 26(4): 1067–95.

Pavarala, Vinod and Kanchan K. Malik. 2007. *Other Voices: Struggle for Community Radio in India*. London, New Delhi, and Thousand Oaks, CA: Sage Publications.

Price, Monroe E. and Stefaan G. Verhulst (eds). 1998. *Broadcasting Reform in India: Media Law from a Global Perspective*. New Delhi: Oxford University Press.

Punathambekar, Aswin. 2011. 'Reality Television and the Making of Mobile Publics: The Case of Indian Idol', in Marwan M.Kraidy and Katherine Sender (eds), *The Politics of Reality Television: Global Perspectives*, pp. 140–60. New York: Routledge.

Rai, Amit S. 2009. *Untimely Bollywood: Globalization and India's New Media Assemblage*. Durham, NC: Duke University Press.

Rajagopal, Arvind. 2001. *Politics after Television: Hindu Nationalism and the Reshaping of the Public in India*. Cambridge: Cambridge University Press.

Robbins, Bruce. 1993. 'Introduction', in Bruce Robbins (ed.), *The Phantom Public Sphere*, pp. vii–xxvi. Minneapolis: University of Minnesota Press.

Sinclair, John. 2005. 'The Indianization of Indian Television', *Flow TV*, 1(12), available at http://flowtv.org/?p=524 (last accessed on 11 January 2011).

Singhal, Arvind and Everett M. Rogers. 1990. 'The *Hum Log* Story in India', in Arvind Singhal and Everett M. Rogers, *Entertainment-Education: A Communication Strategy for Social Change*, pp. 73–104. New Jersey: Lawrence Erlbaum Associates.

Sundaram, Ravi. 2000. 'Beyond the Nationalist Panopticon: The Experience of Cyberpublics in India', in John Caldwell (ed.), *Electronic Media and Technoculture*, pp. 270–94. New Brunswick, New Jersey: Rutgers University Press.

Thussu, Daya Kishan. 2007. 'Indian Infotainment: The Bollywoodization of Indian TV News', in Daya Kishan Thussu (ed), *News as Entertainment: The Rise of Global Infotainment*. London: Sage Publications.

Weintraub, Jeff. 1997a. 'The Theory and Politics of the Public/Private Distinction', in Jeff Weintraub and Krishan Kumar (eds), *Public and Private in Thought and Practice: Perspectives on a Grand Dichotomy*. Chicago: University of Chicago Press.

———. 1997b. 'Public/Private: The Limitations of a Grand Dichotomy', *The Responsive Community*, 7(2): 13–24.

White, Mimi. 2003. 'Flows and Other Close Encounters with Television', in Lisa Parks and Shanti Kumar (eds), *Planet TV: A Global Television Reader*, pp. 94–110. New York: New York University Press.

Williams, Raymond. 1974. *Television, Technology and Cultural Form*. Glasgow, Scotland: Fontana Press.

5

From Clients to Consumers

The Missing Citizens among the Indian Television Audience

Dipankar Sinha

History, it has been philosophically stated by Marx, repeats itself
first as a tragedy and then as a farce. However, insofar as the
Indian television's tryst with the audience is concerned, it is a
classic case of cohabitation of both tragedy and farce. This chapter
will explore this contention from the vantage point of two prime
televisual regimes: (a) the days of the state-guided development
path during which Doordarshan (hereafter, DD) used to enjoy the
monopoly over electronic broadcasting; and (b) the post-liberalization era of market-led development, in which the emergence
of the multiple corporate television channels coincided with the
steady decline of DD's monopoly and credibility. The 'seed' of the
chapter lies in an unfashionable reverse take which steers clear
of the reductionist tendency, to look for the hidden transcript
of continuity in the two aforementioned periods—as opposed to
the much-focused discontinuities, contradictions, and ruptures
that are highlighted by posing them in binary terms. The main
thrust of the argument here is that in the DD-guided televisual

regime,[1] indoctrination has been the communicative foundation of the process of state-sponsored nation building, with the audience treated as 'clients' of the interventionist state, while in the private commercial channel-driven era, hyper-entertainment has been the major instrument of the competition for market share, with the audience treated as consumers. In both cases, I shall argue, the fate of the audience remains largely the same, with their status as *citizens with potential for application of critical faculty*, to be explained subsequently, being the blind spot. Here, I would like to submit that both the periods, marked by different strategies and practices, have a commonality in the sense that they do not leave much space for exercising the critical insight of the audience.

In terms of its methodological orientation, the chapter rests on two hypotheses: first, in general, the television's encoding—in which we include both framing and representations—by itself gives rise to a deep-rooted, multidimensional, and complex process of *reviewing*. The communicative milieu in general, and the televisual regime in particular, in a functioning democratic polity like that of India demands that the audience would indulge in not just passive 'viewing' but also in active *reviewing*. Reviewing being an act of critical evaluation, in hindsight, requires shift from the subject position of viewing. Second, in the specific Indian context, beneath many instances of sweeping changes in the broader political economy during the transition from the state-controlled era to the market-dominated era, lies what I have already described as the hidden transcript of continuity insofar as the encoding of two prime motivated practices—indoctrination and entertainment—and the related issues, messages, and images of the respective televisual regimes, are concerned.

The Audience: Being There without Really Being There

Audience have been an abiding topic of interest with shifting viewpoints (Allor 1988; Webster and Phalen 1994) in media studies in general, and in television studies in particular, for a fairly long time. The stage was set by Morley's (1980) classic empirical study in which the concept of 'preferred reading' established itself as a balancing point of reference by taking into account both the intended closure of the text and the multiple meaning generation by the audience. In terms of its etymological–epistemological roots, the concept of audience has an aural core in the sense of the act of 'listening' of a collective. But with the coming of the audiovisual media, the parameters have been drastically stretched, and the media analysts have gone beyond the set epistemological boundary to indulge in intense debates with two extreme nodes—the *audience are the king* and the *end of the audience*. In the process, the concept has gained lot of elasticity along with an increasing search for the X-factor for better understanding of their diverse responses, which cannot be adequately explained either in reductionist or in ultra-relativist terms. A recent effort in the specific context of the Indian media and democracy expands the conceptual parameters further by highlighting the interpretive practices of the audience–citizens (see Harindranath 2009). It is to be added here that the complex and provocative concept of audience–citizens is witnessing new twists and turns in the era of liberalization in India—a point that would be reflected in the discussion that follows.

When it concerns the study of media effects, the scholars have been broadly divided over the 'Drip Model', in which the television is supposed to influence the audience drip-by-drip, and the 'Drench Model', in which the audience are drenched in terms of influence-seeking representations. Television's influence on the audience has produced a variety of interpretations. In one of the earliest studies, Blumler (1970) mentions the 'interaction effect' of television which, in a 'domestic setting', would have lot of influence on the generally uninterested audience. In contrast, Klapp (1982) relates the constant exposure of the audience to 'museum fatigue'—fatigue caused by creeping boredom after an initial period of excitement and exposure. The latter argument would thus imply that there is a sort of law of diminishing utility operating in terms of the acceptance of media representations by the audience. But as things stand now, the scale of audience research weighs heavily towards the audience's power and agency. Thus, the constructionist approach in media studies shows how the television audience invariably *bring in themselves* in the process of mediation. But whatever may be the contentions of the media theorists, those who run the mainstream media are themselves seem to be convinced about the near-totalistic power of the media over the audience, though their methods and strategies may widely differ in both time and space.

The discussion merits a clarification here: I do not intend to subscribe to the basic tenet of the Frankfurt School, especially the 'culture industries' argument which forecloses 'negotiated' and/or 'oppositional' audience *decoding. Decoding* is Stuart Hall's (1974) well-known concept, which has been expanded by later studies (Lewis 1991; Philo 1990). The thrust here, on the one hand, is on the modes of encoding which, from the encoder's end, is meant to ensure uncritical 'dominant' decoding by the forces owning and managing television broadcasting. On the other hand, the thrust incorporates the 'active audience' thesis in which the potential power and agency of the audience make them the active inter- preters of mediated texts. The 'starting point' here is that while it is important to continue the debate on the nature and extent of the 'audience power' *a la* Fiske (1987, 1989), in India, it is no less important to explore whether television provides *opportunity* to exercise that power.

The process of indoctrination of the *audience as clients* has traditionally been associated with the state-owned media and in the analyses of the Western scholars, especially during the Cold War. In this case, the obvious target was the erstwhile socialist states, especially, the Soviet Union. The indoctrination of televi- sion audience in a democracy like India, as we shall note subse- quently, has concerned the scholars much later and on a much lesser scale. On the other hand, television entertainment as a theme has occupied the mind of the media scholars for long. As Fowles (1992: iv) writes, 'It is entertainment of the highest appeal, some of it resurrected from the vaults for past network hits to fill empty channels'. In the conventional wisdom of televi- sion literature—at least till the expanding influence of postmod- ern perspective—entertainment has been regarded as a collective enterprise. Thus, Mandelsohn and Spetnagel (1980: 13) would remind us: 'To understand entertainment properly, it is important not to focus solely on individualistic pleasure-seeking behaviour. Entertainment phenomena do not take place in vacuum...Rather, entertainment occurs within a context of complex interactions that involve institutions, social norms, group behaviour, and traditions—all of which can be considered to compose a social enterprise.' However, the moot point is whether the audience look for *only* entertainment from television or it is the television proprietors and managers who deliberately and tactically make the 'audience demand' synonymous to entertainment. There are scholars, such as Tannenbaum, who subscribe to the former perspective. As Tannenbaum (1980: 1) argues: 'Deliberate exposure of most people to television is motivated less to seek information, as such, but in search of something generally referred to as "entertainment"'. Thus, Tannenbaum, one of the pioneers in the television audience studies, puts the onus on the audience themselves. But there are a number of scholars who would prefer to focus on the latter perspective, and quite a few of them are seeking to privilege what has now come to be known as the audience autonomy.

In the context of a developing society like India, it can be argued that notwithstanding the discriminatory media scenario, marked by severely emaciated public service broadcasting, existence of scattered and weak community media, and a pathetic indifference to the potential of folk and other kinds of non-mainstream media, the power of the audience-cum-citizens to communicate cannot be dismissed as non-existent; nor can it be treated as a 'lost case'. In fact, the audience's role as citizens, both actual and potential, is to be located within the broad framework of the freedom of expression, which is constitutionally sanctioned in India. The point holds true despite the fact that the audience in actuality may not be constructing frames of understanding from television in a conscious manner to turn into citizens. At this juncture, it is important to note that here we use the term 'audience' more broadly, in the sense of the public, who go through the simultaneous acts of reading, viewing, and listening vis-à-vis the specific genre of media. The term 'citizens' is being used more specifically to refer to those who are concerned with *issues of common concern*, which pervade the society, economy, and polity. But the two are not mutually exclusive terms. While the 'audience' may have their share of opinions, perspectives, intelligence, and sense of judgement, it is only when there is an urge to utilize these traits to deliberate on the issues of common concern, and to participate in the process of decision and policymaking, that at least a section of them turn into 'citizens'. The audience, in our view, ought to have the scope and space to consciously acquire, develop, and disseminate opinions and attitudes to gain the status of citizens. Thus, there is a normative core in our position. More specifically for this purpose, what the audience need are the channels to express their opinions and attitudes in adequate ways, which in turn calls for a democratized access to media. Then again, as I have argued elsewhere (Sinha 1997), the very recognition of these attributes is a long struggle because both the state and the market converge in contracting space for public communication in India—supposedly the world's largest democracy. The roots of such 'convergence' can be found in the refusal of both the state and the market to give due recognition to the notion of citizenship—beyond its legalities—as a civic practice, as an identity, and as a space for the exercise of choice and (political) rights (Amaya 2009). This chapter will, however, elaborate the point in the specific context of television in India.

Doordarshan Era: The Indoctrination Formula

In the DD era, one could find the intimately connected processes of the encoding of mechanistic, technocratic messages on the one hand, and their intended effect on audience, on the other hand. The DD's audience, it needs to be asserted here, were on a broader scale the 'clients' of an interventionist state which, after having emerged at the end of an oppressive and discriminatory colonial rule, was supposed to take the country to the high point of 'modernization' and 'development'. The media in postcolonial India was destined to be governed by a complex mix of patrimony, contradictions, ambivalence, and paradoxes. In its media scenario, the then newly independent India inherited the typical British tradition of private commercial press and the government-owned broadcasting. While the Indian rulers could do nothing to change the leverage of the private entrepreneurs in the press arena, the broadcasting arena was an 'appropriate' site for exerting their control. The tradition of control of broadcasting in course of time became too steeped in the Indian soil to be overturned, as it suited the interests of the Indian rulers for generations and across the political spectrum. For them, the reorganization of the then prevalent media scenario was a 'non-priority item' in the agenda of governance—an item which could at best be afforded lip-service. One needs to remember that in post-colonial India, DD has been largely governed by the utterly colonial and archaic The Indian Telegraph Act (1885). The Act had been introduced by the colonial rulers in their own interest, with little regard to the freedom of expression of the subjects.

Indeed, in independent India, there was the official recognition of the media as Fourth Estate and the freedom of speech and expression was valued beyond its constitutional legitimacy as 'fundamental living value'. But in effect, the All India Radio (AIR) from the very dawn of independence, and the DD since its emergence in 1959 and its subsequent expansion, came to be considered as essential instruments for defining and propagating the definition and function of the then newly formed nation-state and its Nehruvian ideals. The idea was that the patron-state, which had by then exerted monopoly control over the gigantic task of development and consequently assumed mammoth responsibility of directing the lives of the citizens, would communicate with the latter through its own media. In this context, B.P. Sanjay (2006) observes: 'The Nehruvian approach towards institutionalizing the Planning Commission and the centralized approach towards development coincided with his belief that plans and programmes have to be carried and communicated to the people'. But the DD was too precious and powerful a medium to be left to do the publicity of the development plans and programmes in a narrow sense. Thus, it became a vehicle for publicity of the grand Nehruvian paradigm of nation building which outlived Nehru (who died in 1964) and continued to have its hegemony in the successive decades. The DD's emergence and its audiovisual potential were too tempting for the Indian rulers bent on educating the audience to mould them into 'citizens' in a mechanistic sense.

Admittedly, there were a number of historical, political, economic, and social factors, too many to be enumerated in this chapter, which had converged to make the Indian state the prime factor in the nation-building process and its main instrument, development. But the solution became the problem itself. It is due to the astonishing degree with which the state unilaterally indulged in nation building by pursuing 'development' that led to its detachment from citizen engagement in these processes. The Indian state, in seeking to live up to the reputation as the 'harbinger' of nation building in general, and development in particular, would attempt to do it through a vertical top-down mode of information dissemination, not paying much heed to the *shared understanding* without which development even as an engineered process of social transformation becomes both elusive and self-defeating.

In this scenario, the DD with its audiovisual power became a medium *par excellence* for the Indian state. It became a vital publicity organ of the long statist nose which would be poked in almost every conceivable segment of development. A number of studies reveal DD's role as a major publicity organ of the omnipresent state (Gupta 1998; Ohm 1999; Page and Crawley 2001). Thus, Ohm (1998: 82) writes: 'The state's definition of Doordarshan has consisted of its central vision: that the future should bring forth an educated, civilised and united citizenship. Long after the proliferation of the private satellite channels DD's "main aim" remained "national integration", inculcating a sense of unity and making people proud that they are Indians.' Ohm's comment is transparent enough to unravel the role–performance of the Indian state as the custodian of development of the society. The omnipotent and omnipresent state, when it came to guiding the destiny of its citizenry, became overwhelmingly self-empowered and the DD was just the medium to make the audience aware of what is being done for them. Interestingly, the first two of the three goals of DD, *to educate* and *to inform*, which are also major instruments of the process of 'nation building', became predominant in a perverted way, and the third, *to entertain*, was relegated to the background. In its patronizing and patrimonial act and tenor, the Indian state not only instilled in DD its own spirit but also deployed enough surveillance to prevent any kind of deviation, however minor.

The point largely holds true even in the days of Prasar Bharati, which has been in existence for more than a decade to provide 'autonomy' to DD and AIR.

In more concrete terms, DD's programmes—for instance, on top-priority items like agricultural development or industrial development—the overwhelming orientation would be towards *top-down flow* of experts' advice by way of 'informing and educating' the target audience—in these cases, the farmers and the factory workers and/or factory owners—about the happenings in the respective fields. Programmes on health would thus overwhelmingly focus and proclaim the virtue of family planning with many other important health-related issues taking a backseat. The technology-based programmes would extol the virtue of technology, with the whole discourse guided by the technocratic assumption that those who resist the technological progress are basically 'laggards'. The so-called scientific programmes would seek to instil 'scientific temper and spirit' in the audience as the foundation of the nation building process. Vertical modes of information dissemination without a *shared context* would leave many vital questions from the 'targeted' audience unaddressed and unanswered. The close-ended monologue would reach its height during the days of the Emergency in the mid-1970s when the constitutionally guaranteed fundamental rights of the Indian citizens remained suspended and the disciplining instruction from the Indian state to its citizens was *work more, talk less*. The DD would be 'overworked' during the Emergency days as it would continue to bombard its audience with the glowing virtues of the 20 Point Programme of the Government of India,[2] which was supposed to catapult the Indians to the 'apex of development'. While the privately owned commercial print media was, at least theoretically, treating the readers as 'citizens with critical faculty', as proved by the government's imposition of censorship on the newspapers, the DD's case was different. Its 'information-generating' programmes would treat the audience more as passive recipients, with gross underestimation of their potential to provide the essential feedback in determining their own development destiny. The programmes, as a result, became mechanistic, unilateral, and often only ritual transmission of messages to the masses. It in turn resulted in cordoning off debates, absence of opposing views, promotion of stereotypical images, and manipulative interpretation of reality to cater to the needs of the regime in power.

On some occasions, the DD involved itself in some grand development-related programmes to reach out to the common people (Khurana and Chaudhary 1993). These endeavours were not to make a departure from the indoctrination paradigm but to lend a human face to the paradigm itself. I can make a brief reference to these ventures. The Satellite Instructional Television Experiment (1975–6), popularly known as the SITE,[3] was a highly publicized 'grand socio-technological experiment' which would broadcast television programmes to 2,400 villages of six Indian states. Second, the Kheda Communication Project (KCP), started in 1976, was a notable step towards decentralized broadcasting. It relied on participatory two-way teleconferencing for development training in model television stations and sought to involve, through training, the representatives of the panchayats (the units of rural local governance), workers of milk cooperatives (Kheda being part of the Amul network), Anganwadi workers, and primary school teachers. It also resorted to the 'campaign mode' which is an essential technique for generating popular awareness, and adopted various formats for this purpose, such as the puppet shows. The KCP also succeeded, to some extent, in bringing the mainstream Indian academic institutions like the Indira Gandhi Open University and non-governmental organizations under its network. No less important, long before the intense debate on the localization of software, it had also stressed on the development of 'local software' to provide appropriate technological support to the people at the grassroots level. The KCP incidentally received the United Nations Educational, Scientific and Cultural Organization (UNESCO) prize for having generated 'rural local effectiveness', but in a pathetic instance of privileging the urban at the cost of the rural, it was terminated in 1985 to facilitate the establishment of a second television channel in Chennai. Third, the Indian Space Research Organization (ISRO)-sponsored Jhabua Development Communication Project was initiated in 1996 in Jhabua, a backward district in Madhya Pradesh, mostly inhabited by the tribals. It rested on communication-support development to promote projects related to watershed management, health (especially of women and children), non-formal and adult education, rural local self-government, and diverse issues falling under the labels 'socio-economic' and 'culture'. Significantly, it emphasized the participation of local people and the greater and consistent use of local language, local skill, and local knowledge. To reiterate, these instances of DD's 'people-centric' ventures, some sort of deviations from the mechanistic indoctrination, failed as they lacked sustaining power, to break new ground in terms of audience perception.

What the requirements of development communication could not achieve, perhaps the imperatives of political communication could. An incisive study by Rajagopal (2001: 3) shows how DD, through the telecast of the Hindu epic *Ramayana* in 1987–8, would open itself up as a 'device' for broader but more subtle political agenda marked by two major components, neoliberalism–globalization and Hindu nationalism, to promote 'popular participation without requiring popular control'. From the vantage point of this chapter, entertainment in an evident way became the perfect conduit of this shift. No wonder, the telecast of *Ramayana* with an incredibly entertaining plot signalled the coming of another televisual era.

Early 1990s: Waves of Change

The 1990s brought in sweeping changes in India with the formal adoption of the liberalization of the Indian economy in July 1991. Significantly, next year, STAR TV made its entry to India, signalling a revolutionary change through transnational broadcasting in the Indian media scenario and in the Indian market composed of the vast Indian middle-class consumers with potential, actual, and growing purchasing power (see Butcher 2003). The Indian state, long habituated to enjoying monopoly power over broadcasting, had to negotiate with the new milieu in policy terms (McDowell 1997; Shields and Muppidi 1996; Thussu 1999). With the phenomenal increase in the quantum and power of the media sector vis-à-vis the market, the Government of India had to initiate several measures to stimulate the growth of the sector by permitting foreign direct investment (FDI) in the film industry in the year 2002 and in the same year, in the news-related print media. In mid-2011, the Government of India raised the limit of the FDI in radio industry to 26 per cent.

The impact of liberalization—characterized as it is by the advent of the market and the virtual retreat of the state from some key domains of governance—was particularly heavy on the state-controlled media, especially DD. All India Radio (AIR), within a few years of liberalization, had already been experiencing a decline in audience and credibility. The print media, with a private entre-preneurial tradition, was not much disturbed except for the occa-sional threat of 'foreign stake' in ownership and entry of the foreign newspapers and journals. In contrast, DD received the heaviest blow immediately to its monopoly. The Supreme Court, in a land-mark judgement in a case in February 1995, ruled that the gov-ernment's monopoly over broadcasting was 'unconstitutional'.[4] While the judgement was not particularly in favour of broadcast-ing being left free and without any control in the hands of the pri-vate media organizations, it did prepare the ground for deregulated broadcasting, setting free the televisual environment for competi-tion. Yet another move, a result of both a long struggle for media autonomy and the landmark judgment mentioned here, was the declaration of the autonomous status of the DD through the estab-lishment of Prasar Bharati (Broadcasting Corporation of India) on 23 November 1997. Beyond such legal–institutional changes, the 1990s also brought profound changes in the broadcasting arena with the entry and quick-paced penetration of the satellite and the cable channels. It was no longer possible to keep the 'satellite invasion' at bay by branding it as 'cultural invasion from the sky'. The DD thus had no other option but to face the ball game in a highly dynamic and extremely competitive environment.

The DD's decline in power and status, precipitated among other factors by the rise of a number of proactive *private* competitors, led to its uneasy negotiation with the new reality. Such uneasi-ness would primarily relate to devising ways and means to strike a balance, on the one hand, between revenue generation which is essential to survive in the newly competitive environment, and on the other hand, the imperatives of public service broadcasting which itself has a complex conceptual mix in combining an abiding concern for preservation and promotion of 'national identity' with the twin tasks of strengthening civil society and facilitating the growth of public sphere. It is noteworthy that the Prasar Bharati Review Committee[5] emphasized the need for the public service broadcaster to strike such a balance in the emerging reality in the interest of the citizens. But DD's long tradition of indoctrination became too preponderant to reorient itself in favour of the 'thinking' citizens despite having the advantage of being the national media with the widest reach and formidable, albeit underutilized, infrastructure.

Corporate Channels: The Entertainment Mantra

There is no doubt that with the steady consolidation of liberalization, the balance in the televisual scenario in India has heavily tilted towards the corporate-controlled channels, leaving DD in a kind of paralytic state. For reasons too obvious to be explained in detail, the for-profit corporate channels are vying with each other to attract the attention of the audience. In this scheme of things, entertainment has been found to be the gateway to win the audience's mind and ensure huge profit. In total contrast to DD, to which revenue generation was secondary, as it was enjoying the patronage of the Government of India, the corporate channels are supposed to survive if only they can generate profit. Notwithstanding the shrill cry of each corporate channel that they are totally committed to 'objective and truthful' news and 'unbiased' views (the latter is perhaps an oxymoron), orienting the audience towards 24×7 entertainment is at the top of their agenda. Ironically, in many cases, few former top officials of DD and many of its more efficient programme executives became the architect of the entertainment overdrive that would be presented to the audience because they joined the corporate media either as proprietors or as high-ranking employees. Interestingly, DD, after getting a grievous blow from the liberalization venture, also tried to resort to entertainment, especially being encouraged by the Mahalik Committee (constituted to scrutinize DD's revenue generation) which, in its report, recommended entertainment orientation for it by way of introducing appropriate film and serial-based programmes in the afternoons and evenings. But the recommendation raised much hue and cry in the Parliament and ultimately, DD did not succeed in competing with the younger, more professional, and more vibrant channels. As fate would have it, DD, which for long never really looked beyond the film song-based programme *Chitrahaar* in entertaining its audience, was criticized heavily for encouraging entertainment. In a significant observation, a report draws our attention to this: '[T]elevision, caught in the cleft-stick of raising resources and filling expanded broadcast time, had abandoned all its social objectives and placed itself at the mercy of advertisers...It is for the television to modulate and moderate programme content. The norms *and methodology...have to be carefully worked out...*' (NAMEDIA 1986: 34; emphasis added).

The points being raised here do not seek to devalue entertainment as such. It is the hyper-entertainment which not only weakens the other core segments of television's power and credibility but also greatly underestimates the critical faculty of the audience. While it is not easy to exactly quantify the 'desirable' degree of audience autonomy, the fact remains that the audience must have the opportunity to take their own decisions and develop their own perspectives while getting informed by the television. The problem is also not so much with the programmes created with the explicit purpose of entertainment: the soap opera or the slapstick comedy variety. The problem lies precisely with the programmes which seek to mix entertainment in basically non-entertainment programmes, such as news. News, notwithstanding some cynical scrutiny it periodically receives in the west as a 'declining source' political information, still retains a fair degree of audience attention in a country like India. The claims of television channels of being 'most newsworthy' and being 'always with the news' substantiate the point. But the search for maximization of profit leads to the maximization of lifestyle entertainment in which even items which are supposed to elicit critical response from the audience-as-citizens are encoded for the audience-as-consumers. This, as Edwards and Cromwell (2006) point out, ends the tall claims of fair performance and professionalism by the 'liberal' media. For McChesney (1999), as the subtitle of his book suggests, it is 'communication politics in dubious times', that too, with not very positive consequences. Quite paradoxically, the market theory and its consumer model rationalize and celebrate 'free press' for its presumed role in creating informed public who can make reasoned civic judgements based on the choice of the best among the alternatives.

Political entertainment has become the new phenomenon to reckon with. With a complex mix of popular culture, civic culture, and television representations, it is giving rise to intense academic scrutiny (Delli Carpini and Williams 2001; Kellner 1990; Jones 2005; Zoonen 2005). Thus, as already mentioned, serious, business-like news is presented in a highly entertaining manner with the verbal and non-verbal communication of the presenter—the words used, the body language, and the tone and gimmicks—matching that of an entertainer, backed up by flashy images and 'appropriate' sound effects. Such approach is based on the assumption that the audience are not much serious about news. As Lance Bennett (2005a: 173) explains:

[n]ews sagas...result in journalistic constructions that transport focal elements of the event out of context into what might be called a news reality frame. A news reality frame is a decontextualized account based on a documented element of an event that becomes journalistically repackaged in a different story frame. The resulting news reality frame blurs the connection between the new reality and its original surrounding context.

But even this process has to be 'normalized'. In this context, Bennett in *News: The Politics of Illusion* (2005b: 47–64), not just identifies factors like personalization, dramatization, and fragmentation, but also refers to normalization as a strategy by which the encoders of news generate familiar images, both moral and empirical, of a normal world to drive bothersome details out of mind.

This very assumption that audience can be manipulated in this manner goes against the loud claims of the television channels that 'we are here to inform' or 'we present the reality that is there'.

Whether the post-liberalization televisual scenario in India ushers in a 'new dawn' in terms of expansion of public space is therefore a moot question. Theoretically, there is greater scope for it because of the expansion of choice through proliferation of television channels. But the point need not be overemphasized. The question is: whose choice is it? One can, in this context, invoke Neil Postman's (1985) inimitable study, *Amusing Ourselves to Death*, in which he elaborates, in the context of television news in the United States (US), how the 'debate-setting' scenario, marked by the circulation of views and counterviews, is being replaced by 'drama-setting of table-talk' and orchestrated talk shows on development. Postman, to add, is a die-hard critic of television, who would go to the extent of arguing that the medium is 'epistemologically incapable' of democratic viewing and is only fated to spell disaster to the critical faculty of the audience and public discourse. He is extremely pessimistic about the contemporary media ambience. It is true that there are media scholars who do not wish to go as far as Postman, and their critique is accompanied by the claim that there is some good (democratic) sense in the media in general, and even in television balancing. But it is also true that a substantial number of media scholars are very sceptical of the television's infotainment strategy which is 'more of entertainment and less of information'. This trend has also given birth and circulation to the concept of tabloid television which has been critiqued (Smith 2009) for presenting or, pretending itself as being on the side of ordinary people, despite being driven by economic interests of media corporations.

On certain occasions, the mainstream Indian media also realizes that infotainment is being pursued not only at the expense of the audience but also at the cost of their own credibility (Matthews 2009). Thus, for instance, the mainstream Bollywood itself has made a film like *Rann*, in which the protagonist, a television baron played by megastar Amitabh Bachchan, suffers from the agony caused by his son's zealous drive for higher target rating points (TRPs) for their channel, by compromising the ethical issues. The contention is dramatized between two television channels, Headlines 24×7 and India 24×7, symbolizing the breakneck speed with which the audience are made the passive recipients of information. The publicity blitz of the film, very revealingly, promotes the following in English, 'Will we ever know the truth?', and in Hindi, 'Aajkal competition zyada, ideas kum' (nowadays competition is greater, ideas scarce). The dialogues include, 'News ko masala banakar becho' (make the news spicy and sell). Yet, these self-critical views are more of exceptions.

The Indian television channels indulge in periodic debates on the role of the media, in which the anchors and panelists engage themselves in 'critical' discussions on 'contentious' issues. Then again, many of such programmes themselves have good dose of entertainment with lot of antics and dramatics, almost with the feel of a reality show, a phenomenon to be discussed subsequently. The Indian audience also have had the chance to see a highly 'entertaining show' during the Mumbai massacre on 26 November 2009. It was a show of 'embedded journalism' with the journalists and camera persons literally risking their lives by vying to be closest to the 'actors'—the terrorists, the hapless victims, the commandos—for live television 'feed'. In the process, the media persons become the 'co-actors' themselves. Media's excessive zeal to disseminate second-by-second information drew sharp criticism from various segments of the civil society, including some media personnel themselves. Music director Vishal Daldani has been the architect of a public interest petition against sensationalization by television channels.[6] In the petition, he contends: 'What they (the televison channels) were broadcasting in the name of the news, were in fact the exact operational procedures, locations, and actions of our anti-insurgency forces! Minute-by-minute!...Is it acceptable to us that what should have been a classified operation, was in fact an open book?' There are other instances too. A media watch organization has argued that in the zeal for 'glamourization' of Arbind Rajkhowa, the militant United Liberation Front of Assam (ULFA) leader from Assam, and his associates, the television channels of north-east India have not only been indifferent to the mass killings of ULFA, they have also been unkind to Rajkhowa's children who were too frequently flashed on the screen as part of the 'soap opera' (Correspondent 2010).

Then again, certain fuzzy issues remain. Right from the days of the Gulf War, which according to many media analysts was the source of the so-called CNNization of the world, on-the-spot television reporting led to a complex mixture of information and entertainment, sometimes too entangled to have a clear distinction. Yet another motivation for such embedded reporting was a sort of patriotic fervour which has its roots in the 'my motherland is under siege' frame of mind. Can television resist the temptation of hyper-entertainment? It is by itself a complex question. Notwithstanding the talk shows and phone-in programmes, television is a non-interactive medium in the sense that the audience have little scope to chip in unless asked to do so. But at the same time, since it is endowed with, to use Gitlin's (2002) phrase, *torrent of images and sound*, television has a 'built-in impulse' to construct a communication environment in which the audience would not feel the lack of interactivity. This is why hyper-entertainment with high market potential becomes television's 'basic ingredient'. Thus, while news should theoretically be based on impersonal, dispassionate, non-dramatic (re)presentation of events, for the corporate television channels, news is also 'good material' for infotainment (Thussu 2008). This is the reason why, as we have in a Bengali channel, the celebrities are brought in to read news. Beneath the events that news is supposed to broadcast lies the strategy of *eventability*,[7] the event/non-event interplay based on extra-journalistic spin and relentless repackaging of information, as epitomized through 'Breaking News', '24×7', 'Alert', the now familiar captions for spectacular short-term *attention bursts*. It hardly matters that they have little connection with the main content of the news being presented in the larger portions of the screen. In this context, one cannot but invoke Baudrillard (1994) for whom *simulacrum*, the excessive semio-realization of the real, takes over the reality by diluting the distinctions between the signifier and the signified. In Baudralliard's hyperbolic language, it leads to *liquidation of all referentials*.

The episodic, individualized, and hyperactive representations are now scaling new heights in the so-called reality shows which throw up a new challenge to media theorists. Mark Poster (2007: 151), in an essay, poses the question that looms over the contemporary analysis vis-à-vis the reality show: 'The brave new world of reality television...introduces a fascinating, novel landscape of mediated popular culture. Are we to take these examples as indications of a significant cultural formation, or simply as an ephemeral passing fad of programmers and audiences alike?' Hyper-entertainment has acquired a new height with shows like *Sacch Ka Saamna* (facing the truth)—the Indian version of the global format, *The Moment of Truth*—in which the participants, in the lure of prize money, reveal the innermost secrets of their life, including the 'dark' secrets. To add spice and flavour to the shows, the relatives, friends, and colleagues of the participant are also paraded before the television camera which seeks to capture their expressions of pain, embarrassment, and self-inflicted humiliation as intimately as possible for the benefit of the audience. An increasing number of Indian television channels are into reality shows based on song or dance competition. Some are specializing in 'real time' selection of brides and grooms, and as if the current birth is not enough, even some are claiming to have been specializing in the secrets of the 'previous birth'. A show like *Emotional Atyachar* (emotional oppression, Bindass channel) engages journalists to lure and seduce unsuspecting boyfriends and girlfriends of the selected participants and the 'recorded truth' is revealed to the participant to display her/his emotional outbursts. In the shows dealing with song and dance, there is the 'high drama' around the moment of expulsion of the participants, marked by the display of excessively high emotion. The shows also *manufacture* disputes among the judges, which are laced with threats and counter-threats. All these are supposed to be part of the prepared script of the director of such shows.

The media theorists are leaving quite a few clues for theorizing reality show. Significant works by scholars, such as Boorstein (1961), with his concept of *pseudo-events*, and Lance Bennett (2005b), with his concept of *third dimension of symbolic space*, have helped the media analysts to understand the mechanics of the reality show as preconceived constructions to take the audience away from the contextual and underlying reality. One columnist has a different take on television's negotiation with the reality in general, with implicit reference to the audience (Desai 2009). As he argues: 'By virtue of being set in continuous time, (television) tends to narrativize reality into stories...What does not fit into these neat stories becomes difficult to deal with. So it matters little if the story changes very frequently, but the world must come to us in little capsules of digestible stories' (Desai 2009: 4). In Poster's view (2007: 175), while on the one hand these shows invite criticism for promoting patriarchy, capitalist ideology, neoliberal market culture, excruciatingly bad taste, deplorable culture of the masses, shameful exploitation, audience manipulation, heterosexual normativity, postmodern imagination, decadent American civilization, relentless barrage of images of beauty, consumer narcissism, boredom, a culture of superficial amusement, and so forth, they, as Poster himself argues, also open up new vistas not in terms of achieving their 'true' identity but in terms of 'exploring personhood in the age of information machines'.

The 'Frying Pan' and the 'Fireplace'

While there is more to the mediated democracy than the above-mentioned modes of indoctrination and entertainment, the preceding discussion elaborates the fate of the television audience in India, who have moved from the 'frying pan' of the state-led indoctrination to the 'fireplace' of the market-sponsored hyper-entertainment. In making such a transition, insofar as the encoding process is concerned, the audience hardly have been able to gain much in terms of reviewing: exercising their critical faculty in receiving and processing relevant information in order to develop their own perspectives and insights about various issues concerning themselves and others. If in the first era, the messages and imageries of 'development' were almost pushed down the throat of the audience in the name of nation building, in the following era, the excessive doses of entertainment, being sold for profit, itself is development incarnate. In both cases, the audience are sought to be detached from civic engagements and public debates on issues of common concern. The only difference is that in the former case, it was done overtly and crudely, dictating the 'passive' clients of the *patron* state, and in the latter, it is being conducted in more covert ways by 'serving' the consumers of the market.

One can, at this point, argue on a technological plane that television is a strong 'deterrent' to the act of reviewing by virtue of being constituted by incessant 'flow' which also ensures that the television's content remains overwhelmingly oriented to the *dictatorship of the present*. But such argument, being highly technocratic, undermines television's role expectation in facilitating public interest and citizen engagement. Furthermore, attributing the underestimation of the critical faculty of the citizens to 'unalterable' technological attributes is to severely undermine the scope of the social space from both sides of the fence: the deliberate act of audience manipulation; and the power of oppositional and negotiated decoding of the televised representations by the audience. It, however, remains a great paradox that Indian television, cutting across differing time and televisual regimes, is contributing to the shrinking space of autonomy. Interestingly, it is being done at a time when various social and political movements in India are advocating greater space for ordinary people in the decision-making process in securing participatory democracy and bottom-up development.

Notes

1. It technically began in 1959 with the coming of state-owned television which assumed the name Doordarshan in 1976, and would last upto the 1980s.

2. The 20-Point Programme was formulated by the Indira Gandhi regime in 1975 for enhancing agricultural and industrial production, improvement in public service, and for eradication of poverty and illiteracy.

3. The SITE was conducted by the Space Application Centre of the Indian Space Research Organization (ISRO) with an agreement with the National Aeronautics and Space Administration (NASA) of the United States (US).

4. The Ministry of Information and Broadcasting, Union of India and Others *versus* the Cricket Association of Bengal. Judgement declared on 9 February 1995.

5. *Prasar Bharati Review Committee Report*, 2000. Ministry of Information and Broadcasting, Government of India, 20 May. In the section on 'The Need for Public Service Broadcasting', the report takes up the

issue of identity of Prasar Bharati, namely, DD and AIR. Acknowledging (Clause 2.1.3) the 'historical reality' that DD happens to be 'one of the largest broadcasting networks in the world', the report calls for a reoriented DD, ready to function as a public service broadcasting 'to strengthen the democratic process by providing information, promoting debate and discussion on all vital issues, and providing a platform for interaction between the common man and the policy maker' (Clause 2.1.7).

6. Vishal Dadlani and Others, 'Petition against Live Coverage of Terror Attacks and Media Sensationalism', available at www.smallchange.in (accessed on 3 January 2010).

7. I wish to thank Professor Sourin Bhattacharya for a helpful discussion on the notion of 'eventability'.

References

Allor, Martin. 1988. 'Relocating the Site of the Audience', *Critical Studies in Mass Communication*, 5(3): 217–33.

Amaya, Hector. 2009. 'Citizenship', in Stephen W. Littlejohn and Karen A. Foss (eds), *Encyclopedia of Communication Theory*, available at www.sageereference.com/communicationtheory/Article_n40.html (accessed on 23 April 2010).

Baudrillard, Jean. 1994. *Simulacra and Simulation*. Michigan: University of Michigan Press.

Bennett, Lance W. 2005a. 'News as Reality TV: Election Coverage and the Democratization of Truth', *Critical Studies in Media Communication*, 22(2): 171–7.

———. 2005b. *News: The Politics of Illusion*. New York: Longman.

Blumler, J.G. 1970. 'The Political Effects of Television', in J.D. Halloran (ed.), *The Political Effects of Television*, pp. 68–104. London: Panther.

Boorstein, D.J. 1961. *The Image: A Guide to Pseudo Events in America*. New York: Athenium.

Butcher, Melissa. 2003. *Transnational Television, Cultural Identity and Change: When STAR Came to India*. New Delhi: Sage Publications.

Correspondent. 2010. 'Militant Handover Morphs into Saas Bahu Soap Opera', available at www.thehoot.org/web/home/story.php?storyid=4281&mod=1&pg=1&sectionId=22&valid=true (accessed on 15 February 2010).

Delli Carpini, M.X. and B.A. Williams. 2001. 'Let Us Infotain You: Politics in the New Media Environment', in W. Lance Bennett and Robert M. Entman (eds), *Mediated Politics: Communication in the Future of Democracy*, pp. 160–81. New York: Cambridge University Press.

Desai, Santosh. 2009. 'Televisionized India: Beyond Small Screen', *The Times of India*, 7 September.

Edwards, David and David Cromwell. 2006. *Guardians of Power: The Myth of the Liberal Media*. London: Pluto Press.

Fiske, John. 1987. *Television Culture*. London and New York: Methuen

―――. 1989. *Reading the Popular*. Boston: Unwin Hyman.

Fowles, Jib. 1992. *Why Viewers Watch: A Reappraisal of Television Effects*. Newbury Park: Sage Publications.

Gitlin, Todd. 2002. *Media Unlimited: How the Torrent of Images and Sounds Overwhelms Our Lives*. New York: Henry Holt.

Gupta, Nilanjana. 1998. *Switching Channels: Ideologies of Television in India*. New Delhi: Oxford University Press.

Hall, Stuart. 1974. 'The Television Discourse: Encoding and Decoding', *Education and Culture*, 25: 8–14.

Harindranath, Ramaswami. 2009. *Audience-Citizens: The Media, Public Knowledge and Interpretive Practice*. New Delhi: Sage Publications.

Jones, Jeffrey P. 2005. *Entertaining Politics: New Political Television and Civic Culture*. New York: Rowman & Littlefield.

Kellner, Douglas. 1990. *Television and the Crisis of Democracy*. Boulder, CO: Westview Press.

Khurana, B.K. and S.V.S. Chaudhary. 1993. 'INSAT Communication through Television', in K. Sadanandan Nair and Shirley A. White (eds), *Perspectives on Development Communication*, pp. 220–50. New Delhi: Sage Publications.

Klapp, Orrin. 1982. 'Meaning Lag in the Information Society', *Journal of Communication*, 32(2): 56–66.

Lewis, Justin Wren. 1991. *The Ideological Octopus: An Exploration of Television and its Audience*. London: Routledge.

Mandelsohn, Harold and H.T. Spetnagel. 1980. 'Entertainment as Social Enterprise', in Percy H. Tannenbaum (ed.), *The Entertainment Functions of Television*. pp. 13–30. New Jersey: Lawrence Erlbaum Associates Inc.

Matthews, Gerard Paul. 2009. 'Infotainment', in Christopher S. Sterling (ed.), *Encyclopedia of Journalism*, available at www.sage-ereference.com/journalism/Article_n196.html (accessed on 23 April 2010).

McChesney, Robert. 1999. *Rich Media, Poor Democracy: Communication Politics in Dubious Times*. Urbana: University of Illinois Press.

McDowell, Stephen D. 1997. 'Globalization and Policy Choice: Television and Audiovisual Services Policies in India', *Media, Culture & Society*, 19(2): 151–72.

Morley, David. 1980. *The Nationwide Audience: Structure and Decoding.* London: British Film Institute.

NAMEDIA. 1986. *A Vision for Indian Television: Report of a Feedback Project on Indian Television: Today and Tomorrow.* New Delhi: Media Foundation of the Non-Aligned.

Ohm, Britta. 1999. 'Doordarshan: Representing the Nation's State', in Christiane Brosius and Melissa Butcher (eds), *Image Journeys: Audio-visual Media and Cultural Change in India*, pp. 69–96. New Delhi: Sage Publications.

Page, David and William Crawley. 2001. *Satellites over South Asia: Broadcasting Culture and the Public Interest.* New Delhi: Sage Publications.

Philo, Greg. 1990. *Seeing and Believing: The Influence of Television.* London: Routledge.

Poster, Mark. 2007. 'Swan's Way: Care of Self in the Hyperreal', *Configurations*, 15(2): 151–75.

Postman, Neil. 1985. *Amusing Ourselves to Death: Public Discourse in the Age of Show Business.* New York: Viking.

Rajagopal, Arvind. 2001. *Politics after Television: Hindu Nationalism and the Reshaping of the Public in India.* Cambridge: Cambridge University Press.

Sanjay, B.P. 2006. 'Communication Policies in the Nehru Era', available at www.thehoot.org/web/home/searchdetail.php?sid=1927&bg=1 (accessed on 14 August 2006).

Shields, P. and S. Muppidi. 1996. 'Integration, the Indian State and STAR TV: Policy and Theory Issues', *Gazette*, 58: 1–24.

Sinha, Dipankar. 1997. 'Public Communication in Information Age: Time for Requiem?', *Economic and Political Weekly*, 32(37): 2326–9.

Smith, Angela. 2009. 'Tabloid Television', in Stephen W. Littlejohn and Karen A. Foss (eds), *Encyclopedia of Journalism*, available at www.sage-ereference.com/journalism/Article_n375.html (accessed on 23 April 2010).

Tannenbaum, Percy H. 1980. 'An Unstructured Introduction to an Amorphous Area', in Percy H. Tannenbaum (ed.), *The Entertainment Functions of Television*, pp. 1–12. New Jersey: Lawrence Erlbaum Associates.

Thussu, D.K. 1999. 'Privatizing the Airwaves: The Impact of Globalization on Broadcasting in India', *Media, Culture & Society*, 21(1): 125–31.

Thussu, D.K. 2008. *News as Entertainment: The Rise of Global Infotainment*. Thousand Oaks, CA: Sage Publications.

Webster, James G. and Patricia Phalen. 1994. 'Victim, Consumer, or Commodity? Audience Models, in Communication Policy', in James Ettema and D. Charles Whitney (eds), *Audience Making: How the Media Create the Audience*, pp. 19–37. Thousand Oaks, CA: Sage Publications.

Zoonen, Liesbet van. 2005. *Entertaining the Citizen: When Politics and Popular Culture Converge*. New York: Rowman & Littlefield.

6

Television News and an Indian Infotainment Sphere

DAYA KISHAN THUSSU

The year 2010 saw the release of two interesting and instructive Hindi films about the state of television news in the world's largest and nosiest democracy. The Amitabh Bachchan-starrer *Rann*, where the veteran actor played with great aplomb the role of the chief executive officer (CEO) of a struggling television news channel, 'India 24/7', in a competitive and commercially driven news sphere, provided a good opportunity to reflect on the role of television news in India. Another film on the theme—though without the gravitas of a Bachchan or the record of a director like Ram Gopal Verma—was *Peepli Live*, a biting satire on representation of extreme poverty on television news in a world of constant battles for eyeballs, also India's official entry to the Oscars. Its debutant director, Anusha Rizvi, herself a former television journalist, was tackling a theme—farmer suicides—which the mainstream media have generally marginalized. While Amitabh's character of Vijay Malik, a crusading and deeply ethical journalist/owner, inspired hope about the potential of television news to expose corruption, empower people, and safeguard democracy, the fate of Natha, the desperate and destitute peasant in *Peepli Live*, invoked cynicism and despondency, to borrow from a song in the latter film, in our '*rang–biranga prajatantra*' (colourful democracy). At the heart of both these films are the questions about the pressures of excessive marketization on broadcast journalism and the emerging nexus between political and media power. In this chapter, I aim to explore how television news in India is facing these challenges. I begin with a brief description of the evolution of television news. Then, I examine its political and economic dimensions and end with some reflections on the globalizing potential of Indian television news.

Growth of Television News

As elsewhere in the world, the rapid liberalization, deregulation, and privatization of media and cultural industries in India, coupled with the increasing availability of digital delivery and distribution mechanisms, has created a new market for 24/7 news. Television in India has demonstrated exponential growth in the past decade: from Doordarshan—a notoriously monotonous and unimaginative state monopoly until 1991—to more than 500 channels, including more than 100 dedicated news networks, making it home to the world's most competitive news arena, catering to a huge increasingly Westernized Indian audience, and indeed South Asian diaspora. In the late 1990s and early 2000s, India's news television sector demonstrated extraordinary growth in the number of dedicated news channels: from one in 1998 to 40 in 2007, to 88 in 2013, most of which were national, but many international in reach, while some catered to the regional markets. Dedicated news networks now operate in a dozen of the 18 state-recognized languages, several of which have large geo-linguistic constituencies, both within the country and among the 25 million-strong Indian diaspora.

Such an extraordinary growth shows how much has changed since the introduction of television in India in 1959 as a means for disseminating government policies, public information, and state propaganda. Its news coverage rarely rose above what critics rightly derided as 'protocol news'. The ostensible aim of Doordarshan was to educate and inform, though it remained a mouthpiece for the government of the day, reflected especially in the unprofessional way its information bureaucrats ran news operations (Mehta 2008). The partial privatization of the airwaves started with the introduction of advertising onto the state broadcaster in the 1970s, followed by sponsored programmes, and received a boost as India opened up to transnational media corporations in the 1990s.

In the initial years of this liberalization, India's own newspaper and publishing houses did not expand into broadcasting, although there was no regulation against private broadcasting. Early converts to the visual media were such newspaper and magazine groups as Living Media, Hindustan Times (Home TV), Business India (BiTV), and Eenadu. Also, Indian broadcasting space has been reconfigured by what has been described as the 'triumph of the liberal model' (Hallin and Mancini, 2004: 251) of media, partly 'because its global influence has been so great and because neo-liberalism and globalization continue to diffuse liberal media structures and ideas' (Hallin and Mancini 2004: 305).

The global media mughal, Rupert Murdoch, was one key harbinger of such ideas in India, transforming the television landscape with his Star TV (Butcher 2003; Thussu, 2007a). Murdoch introduced the concept of 24/7 news to India when, in 1998, he launched an English-language channel, Star News, to coincide with the national elections that year. Since Indian broadcasting regulation prohibited majority ownership of news channels by foreign companies, Star commissioned New Delhi Television (NDTV) to provide news material, including its presentation and packaging. This was a mutually beneficial partnership: NDTV could reach the homes of affluent Indians as well as diasporic audience through the Star platform, while Star could benefit from gravitas of a serious news channel. However, the Star–NDTV contract came to an end in 2003 and NDTV decided to go alone, launching two channels, the English-language NDTV 24×7 and NDTV India, in Hindi. After the split, Star entered a joint venture with *Anandabazar Patrika* to comply with government ownership regulations for news channels. In this new stage of its indigenization, Murdoch's news network transformed itself into a Hindi-only channel to widen its appeal, as Hindi is the language spoken by the largest numbers of Indians (Butcher 2003; Thussu 2007a, 2007b).

Further deregulation during the 2000s transformed the television industry: by 2011, more than 500 digital channels were operating, including some joint ventures with international broadcasters, reaching more than 500 million television viewers (Federation of Indian Chamber of Commerce and Industry [FICCI]–KPMG 2010; Kohli-Khandekar 2010; Mehta 2008; Thomas 2010). This unprecedented growth has been spurred on by an increase in advertising revenue as Western-based media conglomerates tap into the growing market of 300 million bourgeoning middle class with enhanced and demonstrable purchasing power and media-induced aspirations to a consumerist lifestyle (FICCI–KPMG 2010; Ganguly-Scrase and Scrase 2008).

Cable and satellite television have increased substantially since their introduction in 1992, growing annually at the rate of 10 per cent: by 2010, cable and satellite households had reached 85 million. The media and entertainment business, one of the fastest-growing industries in India, is projected to reach nearly $23 billion by 2012, according to the 2007 report, *Indian Entertainment and Media Industry—A Growth Story Unfolds*, prepared by PricewaterhouseCoopers (PwC) for the FICCI. The Indian television broadcasting market is projected to reach nearly $12 billion by 2012 (FICCI 2007).

While television news outlets have proliferated in such a liberalized and privatized new economy, the growing competition for audiences, and crucially, advertising revenue, has intensified. Not dissimilar to trends in the United States (US), the growing commercialization of television news has forced broadcast journalists and television producers in India to recognize the need to make news entertaining. They borrow and adapt ideas from entertainment and adopt an informal style with an emphasis on personalities, storytelling, and spectacle (Gitlin 2002; Hamilton 2003). Like the US, where such moves have been reinforced by the takeover of news networks by huge media corporations whose primary interest is in the entertainment business—notable examples include Viacom–Paramount (CBS News); Disney (ABC News); Time–Warner (CNN), and News Corporation (Fox News)—in India too, media conglomerates who make profit in entertainment industry have investments in news networks. Such ownership structures can be reflected in the type of stories—about celebrities from the world of entertainment and sport, for example—that receive prominence on news, thus strengthening corporate synergies. In the process, symbiotic relationships between the news and new forms of current affairs and factual entertainment genres such as reality TV have developed, blurring the boundaries between news, documentary, and entertainment (Kraidy and Sender 2011). Such hybrid programming feeds into and benefits from the 24/7 news cycle: providing a feast of visually arresting, emotionally charged infotainment which sustains ratings and keeps production costs low. The growing global popularity of such infotainment-driven programming indicates the success of this formula (Thussu 2007a). In India, as one commentator has noted, entertainment and infotainment dominate the 'ABC of Media—Advertising, Bollywood and Corporate Power' (Sainath 2010).

Television News as Bollywoodized Entertainment

In a fiercely competitive and crowded market such as the one in India, news networks are under constant pressure to raise their target rating points (TRPs) and acquire new programming to ensure regular stream of advertising revenue. As cross-media ownership rules are relaxed, there is greater trend towards concentration of media power: non-media groups such as Tata, Adlabs, and Bharti have invested heavily in television and telecommunication. Increasingly large companies are present across various segments of the media and entertainment world: new 'media conglomerates', drawing their inspiration from the US model, are in the making.

The all-news channels in India are just over a decade in existence but even in their early formative years, their pro-business agenda was well formed. There is a tendency to make news entertaining which, in the context of India, means drawing on Bollywood or Bollywoodized content. Such infotainment fare is now common practice on news networks which regularly broadcast 'exclusive' stories about the supernatural and the bizarre as examples of compelling television. There is a noticeable change in style and content, away from a considered, professional approach to a flashier and visually more dynamic presentation; the emphasis seems to be not on the journalistic skills of news anchors and reporters but on how they look on camera, with style taking precedence over substance. Across the channels, an informal, entertaining schedule is created to increase the audience base, ratings and revenues delivering programmes—sports, entertainment, and lifestyle—have increased, while news and analysis have shown a corresponding decline.

The Spectacle of Cinema and Sport

The notion of the spectacle (DeBord 1977) is relevant in the context of the changing contours of television news in India where one can detect an apparent obsession with celebrity culture, which in India centres on Bollywood, the world's largest film factory (Nayar 2009). As elsewhere in the world, ratings-driven television news in India is forcing journalists and news executives to go for the safety of the soft news option. The power of Bollywood to sell television news is illustrated by the way Rupert Murdoch's entertainment channel, Star Plus, employed Amitabh Bachchan to host *Kaun Banega Crorepati*, an Indian version of the British game show, *Who Wants to be a Millionaire?*, giving its launch in 2000 extensive coverage on its sister channel, Star News, and in the process, dramatically changing Murdoch's fortunes in India (Thussu 2007b).

Most news networks broadcast regular programmes about the glamour and glitz of Bollywood. Star News has a daily programme, *Khabar Filmi Hai* (the news about cinema); Zee News runs a daily bulletin called *Bollywood News*; while NDTV 24×7 has a regular programme, *Picture This*—which used to only review Hindi films but as NDTV has started to have business partners from the US, Hollywood fare has become a regular part of the programme. Such coverage also features in the main bulletins on news channels and it is not unusual, for example, to see Bollywood film music used as a backdrop for news stories. In programmes such as *The making of...*, a particular (usually big budget) film or a song sequence is routinely broadcast on news channels. When a new big budget film is released, it is invariably headline news: television becomes the battleground for marketing and promotion, with endless speculation about how a particular film might do at the box office. Most channels also run interviews with stars of the film, not only within the entertainment segment but as the main news item.

Closely associated with such Bollywoodization of news is the lifestyle segment of news channels across the board, with such examples as *Nightout* (on NDTV 24×7) and *After Hours* (on Zee TV) regularly broadcasting from the glitterati party scene. A striking example of Bollywood taking over the airwaves was to be witnessed in 2009 when nearly 16 million people watched the finale of *Rakhi Ka Swayamvar* (Rakhi's wedding), a reality show featuring Bollywood starlet Rakhi Sawant. Broadcast on NDTV Imagine, at the time part of the NDTV Group and managed by the well-known Bollywood director Karan Johar, the 27-part reality show, in which Sawant got engaged to a Canadian citizen of Indian origin from among 16 suitors, received exceptional coverage (Sinha 2009). Other popular and cloned reality shows such as *Bigg Boss* (on Viacom-owned entertainment channel Colors), three series of which have been hosted by Bollywood stars (Shilpa Shetty, Amitabh Bachchan, and Salman Khan) and *Fear Factor—Khataron Ke Khiladi* (on Colors, first series hosted by Priyanka Chopra and second by Akshay Kumar) are indicative of how Bollywoodized content is sweeping the television screens. Most of these reality shows receive extensive coverage in the news media as they translate well for ratings. On the final day of the Sawant's show, for example, Star News, among others, devoted a half-hour programme to the story. A day before the final episode was to be aired, NDTV 24×7 featured Sawant in its weekly debating programme, *The Big Fight*, which discussed among other things, whether reality TV was eroding Indian culture.

Such synergies between Bollywood and broadcast news are not unusual. The company that ran the Hindi news channel, Sahara Samay Rashtriya, is also involved in Bollywood film production and operated a film-based channel called Filmy. Zee News is part of one of India's largest infotainment conglomerates, with extensive interests in the entertainment industry: apart from general interest Zee TV, it also runs dedicated channels, Zee Cinema and Zing (music channel), both Bollywood oriented. TV-18, India's leading content provider and broadcaster of business, consumer, and general news, operating since 1993, has also made forays into film production with Studio 18. Times Television Network has launched its English movie channel, Movies Now—India's first high-definition movie channel. Star News regularly and unabashedly promotes the serials being shown on its sister channel, Star Plus, with such spin off programmes as *Saas, Bahu aur Saazish* (a magazine show about the 'lives' of the characters in the soap operas). As if there were a scarcity of celebrity news on television, there is now a dedicated Bollywood news channel called E24.

Reporting of crime and corruption too has a Bollywoodized aesthetics and attitude in television news (Nanda 2006; Thussu 2007a). One example of the 'Bollywoodization' of news as spectacle was in evidence in the most dramatic way in the way news channels covered the terrorist attacks in Mumbai on 26 November 2008 and the subsequent 60 hours of hostage-taking chaos carried out by a group of terrorists. News networks discussed in detail how Bollywood bloggers such as Amitabh Bachchan and Aamir Khan posted their reflections on the attacks and their aftermath. Even sober networks such as NDTV 24x7 could not resist the temptation of Bollywood: its flagship weekly programme, *We the People*, telecast on 30 November, just hours after the end of anti-terrorist operations in Mumbai, was dominated by a panel comprising Bollywood personalities.

Sport, in particular cricket, is another major source of spectacle for television news, as news networks recognize the primacy of this colonial game in India's popular culture, having become the most important sport among Indians, cutting across class, language, and even gender barriers and second only to Bollywood in its popularity. As one recent study of television news in India notes, 'the Indian television news industry has consciously ridden on the shoulders of cricket' (Mehta 2008: 197). Television has transformed cricket from a gentleman's game to a roaring business, dependent on corporate sponsorship. The live broadcast of cricket, especially the one-day matches and 20/20 games, has been turned into a visual extravaganza. Cricket-related stories appear almost daily on all major networks—and not just on sports news. These include details of private lives of cricketing stars as well as regular narratives on their expensive lifestyles.

When international matches are underway, news networks employ ex-cricketers and commentators to dissect the day's sporting action, with active participation by audiences, naming and shaming the worst-performing players on a particular day. Interactivity is central to such infotainment, as compelling coverage can lead to channel loyalty in the long run and at the same time, provide a new revenue stream for privatized telecom networks, which benefit from this convergence (Majumdar 2009; Mehta 2008).

Cricket coverage also gets good play on such elite channels as NDTV 24×7 which runs two popular daily programmes: *Sports 24×7* and *Sports Unlimited*. The channel's *Newsnight 20/20* claims that it has borrowed the format of the popular 20/20 cricket tournaments, covering 20 stories in 20 minutes, leaving the rest of 10 minutes in the half-hour segment for advertisements.

The entertainment experience of 20/20 tournaments is 'enhanced' by the presence of dancing cheerleader girls at key points of the game as well as Bollywood stars. The 20/20 format of the game has been given a new lease of life with the establishment of the Indian Premier League (IPL), modelled after the highly commodified British soccer industry, as well as American baseball leagues. What distinguishes the Indian experiment is the dominance of Bollywood stars in this new form of televised entertainment. They not only own teams but also act as brand ambassadors for the sponsors of the tournament. Bollywood superstar Shah Rukh Khan, for example, owns Kolkata Knight Riders, which has a brand value of $22 million, while Shilpa Shetty owns Rajasthan Royals, the team which won the inaugural championship in 2008. Another Bollywood star, Priety Zinta, owns Kings XI Punjab. A United Kingdom (UK)-based consultancy firm, Brand Finance, has valued the IPL brand at $2 billion (Indiantelevision.com 2009). The second season of the IPL, which took place in South Africa in 2009, drew in 122 million viewers worldwide with UFO Moviez, a digital satellite cinema network, screening its matches live in cinemas in

India as well as in select cinemas in the US and in the Middle East. Televising sports is big business: sports genre earned around 15 billion rupees of advertising income in 2010, with cricket accounting for as much as 80 per cent, according to industry estimates.

The mixture of Bollywood glitz and big business has made IPL a major media product, with the 'auctioning' of players being covered live by national television news. Nearly $63 million was spent on purchasing 127 players for 2011 IPL, with Gautam Gambhir being the highest paid player, 'bought' by Kolkata Knight Riders for $2.4 million. One commentator called this practice, 'the obscene auctioning of cricketers for equally obscene amounts of money', comparing such commodification to the 'gladiator' metaphor as being 'redolent of slave trading in ancient Rome, sometimes to fight in gladiatorial contests' (Datta-Ray 2011). The arena for such fights is often the television screen—no wonder Multi Screen Media (formerly known as Sony Entertainment Television India), the official broadcaster of cricket's hottest property which bought exclusive rights for telecasting the IPL for the next nine years for nearly 50 billion rupees, was expecting advertising revenue in the range of 10 billion rupees from the fourth series of the IPL in 2011. In 2010, Zee Network acquired a stake in Ten Sports and launched Ten Cricket—a dedicated 24-hour cricket channel.

Televising Politics: Populism and Partisan News

If excessive marketization has contributed to privileging sport and celebrity spectacle in the news, the shift from bureaucrats to marketing executives has also influenced the politics of television news. News is increasingly shrill, bipartisan, and noisy. One commentator described the phenomenon as 'news theatre' (Ninan 2010). During Doordarshan's monopoly of broadcast news, news on television was considered little more than the government's view of the day's events, with only primary definers of news—mostly politicians and other elite groups—dominating the discourse. In terms of presentation and style, the news was bland and bureaucratic: audience interest did not matter, as there was no competition with private television. With the privatization of television and growing importance of regional parties in national politics, many more groups—distinct in terms of language, region, politics, and economic networks—have invested in television news, making it multi-vocal and multilayered in terms of audiences and production values. Zee TV, India's pioneering television conglomerate, was the first network to launch a 24-hour Hindi news channel in India—Zee News. This gave the channel an important voice in the politically crucial Hindi heartland of India, where it had to compete in later years with Aaj Tak, part of the India Today Group and the most successful new channel since its inception in 2000, and Murdoch-owned Star News, among others. In 2010, it launched Zee Salaam—India's first Urdu infotainment channel.

In southern India, the relationship between popular media and politics is well entrenched. The Telugu megastar, N.T. Rama Rao, was catapulted to the post of the Chief Minister in Andhra Pradesh in the 1980s, while M.G. Ramachandran (who had the distinction of being the first film actor in India to be elected chief minister of a state in 1977) and Jayalalithaa Jayaram, both major presences in the Tamil cinema, have had very important political roles—as chief ministers of Tamil Nadu as well as power brokers on the national political scene. In Tamil Nadu, Jaya TV often takes a pro-All India Anna Dravida Munnetra Kazhagam (AIADMK) stand, while Sun TV, owned by Dayanidhi Maran, supports the Dravida Munnetra Kazhagam (DMK) party (Ranganathan 2008). In Andhra Pradesh, there were eight round-the-clock news channels operating in 2010. Many of these have clearly defined political affiliations: Eenadu TV, for example, part of the largest circulated Telugu daily group, extends strong support to the Telugu Desam Party and opposes the Congress Party, while former Chief Minister Y.S. Rajasekhar Reddy's family members own Saakshi TV. In Karnataka, TV9 Kannada remains a major voice, while in Kerala, Kairali TV has pro-communist news agenda and Jai Hind TV has backing from the Congress-led United Democratic Front. In Bengal, the competition between Star Ananda, a 24-hour Bengali news service, and 24 Ghanta, part of Zee Network, is growing, while in Maharashtra, Star Majha has to compete with IBN–Lokmat TV. These channels have become integral part of political communication in their respective states, increasingly being sought after by politicians and their image managers.

This interest in news may reflect not only the low baseline from which it started but also the popularity and diversity of debate in a complex political scene and the 'argumentative' nature of Indians (Sen 2005). News in Hindi and in other major languages in India has a much wider audience base than the elite English-language networks (Ninan 2007). The new visibility of television news has brought new actors on to the national arena as well as influencing the actions of those being filmed. The way television news covered the conflict in Kargil in 1999—India's first televised war (Thussu 2002)—and the communal violence in Gujarat in 2002, the first major riots of the 24-hour television age (Jain 2010), is indicative of the power of visuals to shape the public agenda.

Politicians and spin doctors are quick to make use of this powerful medium for ideological proselytization. It is interesting to note that during the 2009 national elections, both the Congress Party and the Bharatiya Janata Party (BJP) launched television campaigns on Hindi news channels, ignoring the English networks. Their respective campaigns—'*Aam aadmi ke badte kadam, har kadam par Bharat buland*' (Congress) and '*Majboot neta, nirnayak sarkar*' (BJP)—were aired on mainstream Hindi channels watched by the much-maligned 'aam aadmi'.

Debasing or Democratizing the Infotainment Sphere?

Is commodification of television news and its growing politicization expanding or eroding the public discourse in the world's largest democracy? Given the symbolic and semiotic power of television news, is infotainment-driven television news contributing to making India a producer and consumer of commodity capitalism?

Given the obsession with Bollywoodized content and its regionalized clones that characterize much of television content, the 'public' aspects of news seem to have been undermined

(Rajagopal 2009; Thomas 2010). The popularization of celebrity-driven and sensationalist news may have made it a more marketable commodity, but this has also defiled the public discourse, increasingly aiming at the lowest common denominator. With a few honourable exceptions, much of television news has almost negligible reporting of rural poverty or regular suicides by small farmers and of developmental issues in general, as they rarely translate into ratings or interest advertisers, on whose support the edifice of a commercial television news is ultimately based. It has been suggested that in a market-driven economy, the media system 'is not only closely linked to the *ideological* dictates of the business-run society, it is also an integral element of the economy' (McChesney 1999: 281; emphasis in original).

The almost total absence of development news on commercial channels, who perceive their viewers little more than as consumers of infotainment, is deeply ironic in a country which was the first in the world to use satellite television for educational and developmental purposes, through its 1975 Satellite Instructional Television Experiment (SITE) programme. The vast satellite network that now fuels India's information technology (IT) powerbase is being deployed largely to create an infotainment sphere.

As most of television news is in private control, what happens to the public aspects of broadcasting in a country where, despite strong economic growth, more than 400 million people live in poverty—the world's largest chunk of poor people in a single country? The infotainment-driven television news has not yet reached large parts of rural and semi-urban India where Doordarshan still reigns supreme, despite severe competition from private news networks, especially in cable and satellite homes. In 2010, Doordarshan was reaching more than 400 million viewers, while DD News, the first and the only terrestrial news channel in the country, had the highest reach into television households.

Though not as bland as during its monopoly days, DD News still lacks the edge and critical dimension that at least some private channels have earned in the past decade of operation. It has been suggested that infotainment is a 'new format for political debate'—such news has 'put at risk the celebrity status of politicians, who now compete with glamorous personalities from private industry and the media' (Rao 2010: 188).

Infotainment has also been deployed to encourage young people to participate in democratic processes. In Jharkhand, voter percentage rose from 51 per cent to 58 per cent after Indian cricket captain, Mahendra Singh Dhoni, 'batted for the cause of democracy', shortly before the 2009 assembly elections in one of the poorest states in India. The campaign, *'Pappu pass ho gaya'*, based on a popular Bollywood song, encouraged young voters to take part in assembly elections in Delhi in the same year. As Cottle and Rai (2008: 77) suggest, 'In a diverse and plural polity such as India, the communicative structures of television news are particularly important in that they variously enable or disable the public elaboration of conflicting interests and identities and, as such, are imbricated in processes of democratic deliberation and display, or its curtailment'.

With all its faults, television news has contributed to uniting the nation, promoting awareness of its history and languages, traditions and festivals, as well as providing information on the state of the nation—albeit from a particular position and perspective. The exponential growth of regional television news has been a boon for language journalists. Indeed, some of the most influential broadcast journalists, particularly in the regional context, are those working for Indian vernacular television. In the past two decades, the extraordinary growth in the news industry has created a new generation of broadcast journalists, increasingly pan-Indian in their perspectives. One indicative example of such initiatives is NDTV-Hindu, launched in 2009, to bridge the north–south media gap and to challenge the Delhi-centric approach of television news.

Networks such as NDTV 24×7 have arguably broadened the public discourse in India, bringing on board, for example, questions about environmental protection and right to information. The network has undertaken campaigns to promote particular causes and generated both revenues and, more importantly, awareness on issues such as protecting India's national animal, the tiger. Such instances may be characterized by what Cottle and Rai (2008: 83) have called 'the campaigning frame', which 'declares the news outlet's stance on a particular issue or cause and typically seeks to galvanize sympathies and support for its intervention, political or otherwise, beyond the world of journalism'.

Indian Television News in the Global Media Sphere

Given the size and scale of the Indian television industry and the globalization of Indian businesses, the Indian version of news has potentially a much bigger audience base, beyond the diasporic one. Unlike those in the Western world, the media and cultural industries in India are growing rapidly (Mehta 2008). International investment is increasing in India's media sector, as cross-media ownership rules are relaxed (Kohli-Khandekar 2010; Thomas 2010). At the same time, Indian media companies are also investing outside national territories. As an industry report noted: 'Aspirations of Indian players to go global and foreign players entering the industry are likely to help the industry target a double digit growth in next five years' (FICCI–KPMG 2010: 15).

In the past two decades, India has become an important source of media products, both indigenous as well as a production base for transnational, largely US-based media conglomerates (Kohli-Khandekar 2010). According to the United Nation' (UN) *Creative Economy Report 2010*, India showed the largest growth in exports of creative goods during 2002–8 (UNCTAD 2010). Indian entertainment and telecommunication corporations are investing in Hollywood productions in an emerging synergy between Hollywood and Bollywood, creating a 'global' Bollywood (Gopal and Moorti 2008; Kavoori and Punathambckar 2008; Rai 2009; United Nations Educational, Scientific and Cultural Organization [UNESCO] 2009). Indian popular music—an integral part of Hindi cinema—is also being globalized in the process (Gera Roy 2010): in 2008, the Times of India Group acquired Britain's Virgin Radio, the first such overseas acquisition by an Indian media corporation. Looking East, Indian television news may also have potential to enter new markets and contribute to a 'Chindian' public sphere (see special-themed issue of the journal, *Global Media and Communication*, 2010).

In the realm of television news, such global news players as CNN have already entrenched themselves in India by entering into partnerships with Indian companies—CNN–IBN, an English news and current affairs channel, launched in 2005, in association with TV-18 Group. The NDTV Group (New Delhi Television) has strategic ties with NBC, while Times Now, owned by the Times of India Group, ran a joint news operation with Reuters between 2006–8.

Some of these channels, especially those broadcasting in English, have a global reach and ambition. The richer members of the Indian diaspora—estimated to have a net worth of $300 billion and annual contribution to the Indian economy valued at up to $10 billion—are tuning in to Indian news channels and online news portals to keep abreast of developments. NDTV had branched into new territories with NDTV Arabia and NDTV Malaysia, while its flagship channel, NDTV 24×7, was available in 2011 in the US (via DirecTV), the UK (BSkyB), the Middle East (Arab Digital Distribution), and southern Africa (Multi-choice Africa).

In parallel with the transformation of broadcasting sector, there has also been a massive expansion in newspaper circulation: according to the World Association of Newspapers (WAN), in 2009, India had 2,337 'paid-for' daily newspapers—the highest number in the world—and it was also the world's largest newspaper market with 110 million copies sold every day (WAN 2010). In FM radio, the growth was extraordinary: the total number of radio stations was estimated to touch 800 by the end of 2011. Though the Internet penetration is very low—merely 7 per cent of India's 1.2 billion population—its use is growing exponentially, demonstrating user growth of 1,520 per cent in the period 2000–10, according to Internet World Stats (2011).

As television news converges with online delivery mechanisms, it is safe to predict that more and more Indian content will circulate on the web. Given the so-called 'demographic dividend'—more than 70 per cent of Indians are below the age of 30 years—it is likely that as their prosperity grows, a sizeable segment of young Indians will be increasingly going online, producing, distributing, and consuming news, especially using their skills in the English language, the vehicle for global communication and commerce.

Will journalists working in such a scenario care enough about the fate of news as public knowledge, the kind of concerns that preoccupied Bachchan's character in *Rann*, or will globalized info-tainment encourage them to exploit the poverty of people who inhabit villages like Peepli, scattered across the country?

References

Butcher, Melissa. 2003. *Transnational Television, Cultural Identity and Change: When STAR Came to India*. New Delhi: Sage.

Cottle, Simon and Mugdha Rai. 2008. 'Television News in India: Mediating Democracy and Difference', *International Communication Gazette*, 70(1): 76–96.

Datta-Ray, Deep. 2011. 'Checking a Double Negative'. *The Times of India*, 15 January, available at http://articles.timesofindia.indiatimes.com/2011-01-15/edit-page/28371221_1_cheerleaders-cricketers-ipl (last accessed on 5 December 2013).

DeBord, Guy. 1977 [1967]. *The Society of the Spectacle*. Detroit: Red and Black. (First published in *La societe du spetacle* by Buchet-Chastel, Paris.)

Federation of Indian Chamber of Commerce and Industry (FICCI). 2007. *The Indian Entertainment and Media Industry—A Growth Story Unfolds*. London: PricewaterhouseCoopers in association with FICCI.

FICCI–KPMG. 2010. *Back in the Spotlight—FICCI–KPMG Indian Media and Entertainment Report, 2010*. Mumbai: FICCI.

Ganguly-Scrase, Ruchira and Timothy Scrase. 2008. *Globalization and the Middle Classes in India: The Social and Cultural Impact of Neoliberal Reforms*. London: Routledge.

Gera Roy, Anjali. 2010. *Bhangra Moves: From Ludhiana to London and Beyond*. London: Ashgate.

Gitlin, Todd. 2002. *Media Unlimited: How the Torrents of Images and Sounds Overwhelms Our Lives*. New York: Metropolitan Books.

Global Media and Communication. 2010. '"Chindia" and Global Communication', Special Issue, 6(3): 243–389.

Gopal, Sangita and Sujata Moorti (eds). 2008. *Global Bollywood: Travels of Hindi Song and Dance*. Minneapolis: University of Minnesota Press.

Hallin, Daniel and Paolo Mancini. 2004. *Comparing Media Systems: Three Models of Media and Politics*. Cambridge: Cambridge University Press.

Hamilton, James. 2003. *All the News That's Fit to Sell: How the Market Transforms Information into News*. Princeton: Princeton University Press.

Indiantelevision.com. 2009. 'SRK's Kolkata Knight Riders is top IPL brand at $22 mn: Study', 21 May, available at http://www.indiantelevision. com/mam/headlines/y2k9/may/maymam81.php (last accessed on 29 September 2009).

Internet World Stats. 2011. 'Asia', available at www.internetworldstats. com/stats3.htm (accessed on 21 January 2011).

Jain, Anuja. 2010. '"Beaming it Live": 24-hour Television News, the Spectator and the Spectacle of the 2002 Gujarat Carnage', *South Asian Popular Culture*, 8(2): 163–79.

Kavoori, Anandam and Aswin Punathambekar (eds). 2008. *Global Bollywood*. New York: New York University Press.

Kohli-Khandekar, Vanita. 2010. *The Indian Media Business*, 3rd edition. New Delhi: Sage.

Kraidy, Marwan and Katherine Sender (eds). 2011. *The Politics of Reality Television: Global Perspectives*. New York: Routledge.

Majumdar, Boria. 2009. 'Opiate of the Masses or One in a Billion: Trying to Unravel an Indian Sporting Mystery', in K. Moti Gokulsing and Wimal Dissanayake (eds), *Popular Culture in a Globalized India*, pp. 239–51. London: Routledge.

McChesney, Robert. 1999. *Rich Media, Poor Democracy—Communication Politics in Dubious Times*. Champaign, IL: University of Illinois Press.

Mehta, Nalin. 2008. *India on Television: How Satellite News Channels Have Changed the Way We Think and Act*. New Delhi: HarperCollins.

Nanda, Vartika, 2006. 'Television Crime Reporting in India', in Uday Sahay (ed.), *Making News: Handbook of the Media in Contemporary India*, pp. 189–200. New Delhi: Oxford University Press.

Nayar, Pramod. 2009. *Seeing Stars: Spectacle, Society and Celebrity Culture*. New Delhi: Sage.

Ninan, Sevanti. 2007. *Headlines from the Heartland: Reinventing the Hindi Public Sphere*. New Delhi: Sage Publications.

———. 2010. 'When News is Entertainment', *The Hindu*, 22 May, available at http://www.thehindu.com/opinion/columns/Sevanti_Ninan/ media-matters-when-news-is-entertainment/article435001.ece (last accessed on 5 December 2013).

Rai, Amit. 2009. *Untimely Bollywood: Globalization and India's New Media Assemblage*. Durham: Duke University Press.

Rajagopal, Arvind (ed.). 2009. *The Indian Public Sphere: Readings in Media History*. New York: Oxford University Press.

Ranganathan, Maya. 2008. 'Give Me a Vote and I Will Give You a TV Set: Television in Tamil Nadu Politics', in Nalin Mehta (ed.), *Television in India: Satellites, Politics and Cultural Change*, pp. 106–23. London: Routledge.

Rao, Ursula. 2010. *News as Culture: Journalistic Practice and the Remaking of Indian Leadership Tradition.* New York: Berghahn Books.

Sainath, Palagummi. 2010. 'How to Feed your Billionaires', *The Hindu*, 17 April, available at http://www.thehindu.com/opinion/lead/how-to-feed-your-billionaires/article399250.ece (last accessed on 5 December 2013).

Sen, Amartya. 2005. *The Argumentative Indian.* London: Penguin.

Sinha, Ashish. 2009. 'Rakhi's Swayamvar Tops Ratings', *Business Standard*, 4 August, available at http://www.business-standard.com/article/companies/rakhi-s-swayamvar-tops-ratings-109080400042_1.html (accessed on 5 December 2013).

Thomas, Pradip. 2010. *The Political Economy of Communications in India: The Good, the Bad and the Ugly.* New Delhi: Sage Publications.

Thussu, Daya Kishan. 2002. 'Managing the Media in an Era of Round-the-clock News: Notes from India's First Tele-war', *Journalism Studies*, 3(2): 203–12.

———. 2007a. *News as Entertainment: The Rise of Global Infotainment.* London: Sage Publications.

———. 2007b. 'The "Murdochization" of News? The Case of Star TV in India', *Media, Culture & Society*, 29(3): 593–611.

United Nations Conference on Trade and Development (UNCTAD). 2010. *Creative Economy Report 2010.* Geneva: UNCTAD.

United Nations Educational, Scientific and Cultural Organization (UNESCO). 2009. *UNESCO World Report Investing in Cultural Diversity and Intercultural Dialogue.* Paris: UNESCO.

World Association of Newspapers (WAN). 2010. 'World Press Trends: Advertising Revenues to Increase, Circulation Relatively Stable', available at http://www.wan-press.org/article18612.html (accessed on 20 August 2010).

7

When Live News was Too Dangerous

The Early History of Satellite TV in India

Nalin Mehta

'Live News is too Dangerous'

One of the many forgotten strands of Doordarshan's response to the growing popularity of satellite channels in 1994 was the then Information and Broadcasting Secretary Bhaskar Ghose's blueprint for an upmarket and informed current affairs channel. With live news and business programming from around the world, this channel, DD 3, was envisaged as a serious attempt to win back the elites who did not watch Doordarshan but had switched to channels like BBC World and STAR Plus. Ghose's plan provided for independent news bulletins outsourced to private producers— instead of Doordarshan's usual propaganda—with live discussions, chat shows, and foreign entertainment programmes. The channel was formally inaugurated in October 1994, but Ghose almost immediately received an urgent missive from Prime Minister Narasimha Rao, who had just one question: was it true that DD 3 would show live current affairs programmes? When Ghose answered in the affirmative, the prime minister immediately ordered it be shut down. His reasoning couldn't have been clearer: '...We cannot have live broadcasts. It is too dangerous.' The fear was that there would be no way of controlling anybody from saying anything against the ruling Congress on live programming (Ghose 2005: 189–90).

Narasimha Rao had been forced to initiate economic liberalization in 1991, the Indian state's first decisive break with the strictures of the 'licence–permit' raj, but this incident revealed the immense reluctance to dismantle the ideology of control and the political impulses that had been its driving force. Control over television was central to and constitutive of the state's self-image—broadcasting's principal objective so far was to 'display and enact government control' (Rajagopal 2001: 78)—and live television threatened to break down the very edifice on which Indian television had been built. This was seen most starkly in the case of news programming and thus, despite having spent nearly Rs 200 million on DD 3, the Rao government shut it down (Ghose 2005: 190–1).

By 1998, however, the first of India's private 24-hour news channels was on the airwaves and by 2009, more than 400 satellite channels were officially broadcasting into Indian homes. Of these, more than a 100 broadcast news in 15 languages and more than 70 are 24-hour news channels in 11 languages.[1] No other country in the world has such a concentration of private news channels and the numbers are a stark illustration of the massive changes in Indian broadcasting, which until the early 1990s had remained a state monopoly.

These upheavals in the nature of Indian television have been accompanied by a simultaneous expansion in its reach and penetration. In 1992, if you divided India's population of 846,388,000 (Registrar General of India 2001) by the total number of television sets in the country,[2] the number of people clustering around a set would have been a little over 26. By 2006, that ratio had come down substantially to just over 10 people per television set, despite a substantial increase in the population.[3] In a little over a decade, the total number of Indian television households tripled to reach an estimated 112 million (National Readership Studies Council 2006). It made India the world's third-largest television market, just behind China and the United States (US) (PricewaterhouseCoopers and FICCI 2005: 36).

How did this massive transformation unfold? The first part of this chapter sketches the early histories of the first Indian private television networks like Zee, Asianet, and NDTV to argue that a unique confluence of technological, political, and economic factors in the 1990s drove the transformative process that led to the battering down of the government's monopoly over television. Satellite television first came to India as the representative of global capitalism, and Indian television entrepreneurs initially piggybacked the foreign networks, often in subordinate positions, because of financial and technological constraints. By the end of the 1990s, however, the growing strength of Indian capitalism after the liberalization of the Indian economy and the forces of what Thomas Friedman has called 'Globalisation 3.0' (Friedman 2007: 10–11, 237)[4] allowed Indian entrepreneurs to level the playing field.

A recurring theme in this story is that though the Indian state, having embarked on economic liberalization, was forced to adapt to satellite television as an agent of global capitalism, it certainly did not give up control over television easily or voluntarily. It simply lost control. When confronted with change, the centralized Nehruvian state transformed into a multilayered patchwork state with its overall coherence greatly reduced. I have examined this in detail elsewhere (Mehta 2008: Chapter 3), through an analysis of Indian broadcast regulations, but it is important to note here that while the expansion of Indian satellite television took place within the framework of economic liberalization, the process was not as smooth as the narrative of reforms suggests. It was far more complex and it would be wrong to see the terrified sentinels of the *babu raj* and private entrepreneurs as closed categories locked in combat with each other. Various governmental departments and individuals often acted in opposition to each other, with some wholeheartedly embracing the change, while others fought it tooth and nail and pursued their own interests. The Nehruvian highway was not replaced with a sign-posted post-reform autobahn but with an open or opaque field, depending on one's perspective, whose boundaries were constantly being tested and pushed.

The second great theme this chapter touches upon is that much like India's 'newspaper revolution' (Jeffrey 2003: xi) that started in the 1970s, and the 'cassette culture' (Manuel 1993) of the 1980s, the availability of privately produced satellite television has meant that 'people discovered new ways to think about themselves and to participate in politics that would have been unthinkable a generation before' (Jeffrey 2003: 1). Operating at the junction of public culture, capitalism, and globalization, satellite news networks have had profound implications for the state, politics, democracy, and identity formation.

Despite all their shortcomings and sensationalism, the emergence of satellite television news networks has enhanced and strengthened deliberative Indian democracy. It is important to clarify here that there is no evidence to show that satellite television has benefited Indian democracy if we understand it in the narrow procedural terms defined by the voting process alone.[5] My claim refers to a broader understanding of democracy as a deliberative process involving larger collaborative processes of decision making, identity, and interest formation, with the media acting as a crucial hinge. Democracy is intimately connected with mechanisms of public discussion and interactive reasoning. Indeed, the new disciplines of social choice theory and public choice theory are connected to ideas of individual values and their impact on decision making (Sen 2002). In this context, Amartya Sen has famously shown that no substantial famine has ever occurred in a country with a democratic form of government and a relatively free press (Sen and Dreze 1989).

'You Cannot Stop the Sun with an Umbrella': Global Capitalism on Satellite

In 1991, new satellite technology enabled the forces of global capitalism to leapfrog over the barriers of the Indian state. The catalyst was Hong Kong-based billionaire Li Ka-Shing's vision of using satellite technology to create a new trans-Asian television market. It propelled his affiliated company, Hutchinson Whampoa, to launch the STAR TV network in 1991 to target the wealthiest 5 per cent of Asia's population, excluding Japan, with its free-to-air television signal (Ghemawat 2005: 1). Broadcasting primarily in English, but also in Mandarin, such a network had not been possible before because of technological constraints, but Hutchinson Whampoa created it by investing in AsiaSat 1, the first private and commercially available satellite covering the Asia-Pacific region.[6] The satellite, for the first time, made it possible to simultaneously access television audiences from Turkey to Japan and STAR began with five 24-hour television channels broadcast to 38 countries stretching across seven time zones from East Asia to the Middle East. Although India had not been its primary target, it was here that its English-language programming mix of entertainment, movies, sports, and news had its first success (see Yoshino and Williamson 1994). Within six months of its launch, Star, beaming out of Hong Kong, found India to be its largest market (Page and Crawley 2001: 77).

STAR TV initially became popular in India because of two reasons: the mushrooming of satellite dishes that could download the signal; and the 1991 Gulf War with CNN's round-the-clock coverage of the conflict. There was tremendous curiosity amongst India's middle classes to watch the war coverage, unlike anything on Doordarshan, and the demand created its own supply. Private entrepreneurs quickly sprang up in housing colonies across India's big cities to cash in, and the method was simple: anybody who saw a potential in the business would simply buy a satellite dish, install it on the roof of their house, and then connect it to neighbouring houses for a small monthly fee. Thus was born the cable operator who was to be the interface between the broadcaster and the Indian viewer.

The business model was virtually fail-safe. A satellite dish, which measured between 8–16 feet in radius, cost about Rs 100,000 in 1990–1. The broadcast signals themselves were free-to-air and once the dish was installed, it could be connected via video cables to thousands of households. The cable operator only had to incur a one-time installation cost and after this, was assured a monthly subscription income from all the clients.[7] This kind of cable business was not new to India. Since the early 1980s, the massification of television had gone hand-in-hand with the video revolution and Indian towns were full of video parlours that ran similar cable services.

Many of the early cable operators were improvised neighbourhood operations; these were not networks, had no legal boundaries, and served purely local needs. This phenomenon quickly spread in the metros (Page and Crawley 2001: 90). Apart from new entrepreneurs, it was many of these video cable operators who, on seeing the massive potential in the satellite business, quickly bought satellite dishes and hooked them up their existing cable networks. This was how one of the first cable operators in Delhi explained his switch from video to satellite channels:

> We had got into this business for the sake of CNN...to show the war to people in every home. We started first with one video film channel...Then when CNN came, we had to show it also...Previously there was only the video channel...The moment there is more than one channel...you need to modulate those channels into different frequencies...then only they work. So a whole lot of modulators started coming in, receivers started coming in.[8]

This example is fairly typical of how the cable industry began, and it grew exponentially.

Soon after taking over STAR in 1993, Rupert Murdoch recognized the importance of these cable operators, calling them 'pirates', but nevertheless 'splendid entrepreneurs...pioneering the market' (Murdoch 1994: 11). Though Murdoch's companies soon launched initiatives to take back control of distribution, his statement captured something of what was happening. The cable operators had successfully become the middlemen between the broadcasters and television viewers. By the mid-1990s, there were an estimated 70,000 cable operators in India.[9]

The problem, of course, was that under the 1885 Telegraph Act, this entire business of downlinking satellite signals and distributing them for a fee was illegal. The state still retained a complete monopoly on all broadcasting and any such activity on Indian soil was an offence. This was why a network like STAR could not organize its own distribution; it could merely broadcast its signal and only make money from advertising.

As small players, the thousands of cable operators filled the gap between the global broadcaster and the viewer by taking up the illegal part of the transaction. Cable held the promise of a new world, it represented the aspirations of a middle class that was no longer content with the frugal control ethic of the Nehruvian state, and it was an opportunity to make money. According to the founding president of the Delhi Cable Operators Association:

> At that point of time whatever we did was illegal. By the Indian Telegraphic Act we could not receive the signals. But the technology had to come, something new was coming. The government was lagging behind so a whole lot of people jumped into it and the government eventually changed the laws. But if we had [hesitated], there would have been no cable perhaps even today.[10]

The question then is why was the cable industry allowed to grow and why did the state not use its authority to crush what was essentially an illegal business conducted absolutely in the open.

Nikhil Sinha (1998: 24) has pointed out that the state's initial response to satellite television took three forms: (a) an open-skies policy by doing little to stop them; (b) a competitive response by revamping Doordarshan and launching satellite channels of its own; and (c) finally, an attempt to exert some control by providing global media companies with Indian satellite platforms like the short-lived partnership between Doordarshan and CNN. A typical reasoning for the initial inaction is posited by P. Upendra, Minister for Information and Broadcasting in 1990:

> A file was put up in the ministry as to how to counter the satellite invasion. What steps should be taken to stop it? I wrote back saying *you cannot stop the sun by holding an umbrella.* The more you try, the more you encourage people to watch. It was better to allow them to operate from India under control rather than allow them to operate from Sri Lanka or some other foreign country. (quoted in Page and Crawley 2001: 266; emphasis added)

As early as June 1989, the government had set up a committee to study various aspects of the establishment of cable television networks, and number of similar committees followed after Star's launch in 1991, but a major reason why the state did not clamp down on illegal cable operators was the political and economic climate of reform.

India's economic reforms began in the middle of a severe macroeconomic crisis that erupted in early 1991. Its basic underlying features had been simmering since the late 1980s but the crisis was brought to a head 'by a steep fall in foreign exchange reserves to about $1 billion [equal to two weeks' imports], sharp downgrading of India's credit rating and a cut-off of foreign private lending' (Joshi 1994: 204; also, see Roy 1993). Faced with a crisis of mounting proportions, the Narasimha Rao government undertook steps to stabilize the economy and embarked on a systemic process of structural reform.[11] This included changes in industrial policy and the removal of barriers like pervasive licensing controls in some industries; the liberalization of trade policy with the dismantling of the complex import control regime; and reforms in taxation and the financial sector, public sector, labour markets, and agriculture.

Yet, many foreign players at that stage were not quite sure about the longevity of this process. Foreign direct investment was flowing in very slowly and by mid-1992, only $240 million had been committed by foreign investors. Foreign investors were still jittery about the government's intentions and only one-third of this amount had been actually invested. In such a situation, Sinha (1998: 24–8) argues, satellite television became the show window of the reforms:

...the Government's attitude towards the new satellite-distributed television services became a barometer of its commitment to the reforms. Any attempt to restrict foreign broadcasters would have been construed as evidence of the government's lack of commitment to opening up the economy. Consequently, the Government chose to ignore the foreign television services, despite complaints from a number of political and social organizations of the cultural threat posed by these services. The Government clearly recognized that television is a highly visible cultural product that functions as the best marketing tool for the liberalization of the Indian economy.

Once committed to utilizing the market as the main engine of growth, the government did not want to restrict the forces represented by STAR and others.

While this narrative of reform and the desire not to interfere with the forces of the market economy is crucial to understanding how the story unfolded, it does not by itself satisfactorily explain the complex process through which the indigenous satellite networks were created. The liberalization of the economy was the backdrop, no doubt, but the big picture was terribly messy. Some institutions of the state actively aided the process, while others lined up in bitter opposition.

An Indian in a Foreign Sky: 'How to Mortgage a Satellite?'

The established print media barons were initially slow to react to satellite television. Many of them certainly had plans to get into television—some already ran television production companies—but the satellite business was new and the road to starting up a channel was fraught with too many hurdles, not least of which was the question of whether this could be a profitable venture. While big media initially remained cautious, the new road was first travelled by a new breed of entrepreneurs. Among the first were Subhash Chandra, who started the Hindi Zee TV, and Sashi Kumar, who launched the first indigenous channel in a regional language, the Malayalam Asianet. They had very different backgrounds but both foresaw the immense possibilities of satellite technology and were the first to realize that money could be made out of it.

Subhash Chandra says that when he first approached STAR to hire one of the 24 transponders on AsiaSat 1 satellite, nobody took him seriously. Company officials thought he was just another private producer and put him on to a junior girl in the programming section. Some of the big print groups had baulked at the annual rental price tag of $5 million but Chandra, whose first offer was between $1.5 million and $2 million, promptly agreed. He did not have that kind of money but, as he explains, 'I figured that if the channel was going to work, it was going to do so even at $5 million' (Agarwal 1994: 76).[12] After raising the resources from British businessman, Sir James Goldsmith, and from Indians living overseas, Chandra signed the deal with STAR on 23 April 1992 and the new channel, Zee TV, began broadcasting six months later in October (Bansal and Carvalho 1998).

In a similar vein, Sashi Kumar at Asianet found it almost impossible initially to raise institutional financing in India. As he explains, no Indian bank at that stage would lend money for satellite channels which were still unproven as profitable business ventures:

> We would go to the chairman of the bank, the board and I would make a presentation...The chairman would tell me, we would like to support you but you know banks in Kerala, particularly, are conservative. We normally lend money against gold. We procure against mortgage of property or security as tangible assets. You are telling us here that there is a transponder up in the sky. *Now if you default on the loan what do we do? Do we take a rocket and go in search of your transponder. And if we do get it, it's a leased transponder. What do we seize? What is the surety? What do you mortgage?*
>
> Secondly, I was making revenue projections saying I was going to achieve 18 crores by the second year...which in hindsight proved correct. [The bank manager] said how do you arrive at these projections because Doordarshan was the only channel then available and its total advertising revenue [in Kerala] in the beginning of the 1990s was in the region of Rs 8 and 10 crores [Rs 80–100 million]. We had no track record or no antecedent to point to in this business.[13]

It was a business that was little understood and Asianet, initially, survived through a private investment by Kumar's uncle who was a trader in Russia. The refusal of the banks to back the venture meant that Asianet initially ran on what Kumar calls 'starvation mode', and this is an indicator of how heavily the dice was loaded against Indian entrepreneurs at that stage. But the forces of global capitalism had opened a window and television was at its cutting edge. Like Murdoch's 'splendid pirates', the cable operators,

Asianet's founders improvised to raise resources because they saw there was a market that could be created.

While larger economic imperatives dictated that no steps were taken to hinder the cable operators at least until 1994, the Ministry of Information and Broadcasting made every attempt to hang on to its monopoly. The legal and technical restraints on private broadcasting from India remained in place and this meant that Indian entrepreneurs had to evolve an ingenious mechanism of circumventing the guidelines.

Designed as a regional-language channel for Kerala, Asianet had to set up its base in Russia.[14] It first started broadcasts from an earth station three hours from Moscow. In order to make the arrangement work, a courier would fly from Chennai to Moscow daily with recorded tapes containing the day's programming. This became even more complicated when Asianet shifted to the US military base at Subic Bay in the Philippines. The military base, which was then in the process of being decommissioned, offered satellite uplinking facilities. The courier's daily journey now had three components; a two-hop flight to Manila, followed by a helicopter ride to Subic Bay, a journey which by road would have taken about 12 hours.[15]

With no sign of the central government's protectionist barriers coming down, help came from Kerala's state government. Asianet, the channel, had been started along with another company called Asianet Satcom, which was to be a cable distribution company for taking its signals to viewers. The big breakthrough came when Sashi Kumar convinced the then Kerala Chief Minister, K. Karunakaran, to get the state-owned Kerala State Electricity Board (KSEB) on his side. Asianet Satcom and KSEB signed a contract which gave the cable company the exclusive right to use electricity poles throughout the state for a period of 10 years for the price of Rs 10 per pole per year. This gave Asianet Satcom a crucial advantage over every other cable operator in the state and opened the pathway to networking the entire state to cable television. Kumar claims he got the idea for multiple use of electricity poles from Ted Turner's cable venture in the US. But more crucially, the deal was a windfall for the television channel because it meant that financial institutions finally began to take Asianet seriously.

The bankers who had been worried about how they could accept a hired satellite as mortgage security could now see a tangible asset against which to advance money. 'I could [now] get institutional finance,' says Kumar, 'and that is when money started flowing'. 'I could also leverage this and sell some stake...so there was a large infusion of funds to run the channel as well, as well as the cable venture.'[16]

The deal allowed Asianet Satcom to become the first state-wide cable network in India, but the chief minister's decision was crucial. Kumar insists that the chief minister himself never exerted editorial pressures, unlike many others in his government, but there is no doubt that the decision translated into a giant leap for Asianet's early financial legitimacy. No other state government followed the example and it is instructive that this deal straddled a grey area in the law. Electricity is a state subject, so while the state government was well within its rights to grant use of electric poles under its purview to a private player, it was still illegal for a private player to receive and transmit satellite signals under the 1885 Indian Telegraph Act. There was clearly a tussle of jurisdiction here and while the state government's lawyers had approved the deal, Sashi Kumar admits that had it been challenged at the time, the decision could have gone either way. In the event, nobody challenged the deal and KSEB recently renewed its arrangement with Asianet.[17]

Asianet's early funding problems also stemmed from the fact that its targeted market of Malayalee speakers was much smaller than that of a Hindi channel like Zee. Like Asianet from the Philippines, Zee was broadcast into India from Star's broadcast hub in Hong Kong, in order to get around the restrictions on broadcasting. It was aimed at India's estimated 400 million Hindi speakers and became an instant magnet for national advertisers who thought it had unlocked the north Indian market. It proved to be a spectacular success, breaking even on the initial investment within nine months of its launch. This was highly unusual in the broadcast business which usually has long gestation periods.

By 1994, Zee claimed a 65 per cent share of the Indian satellite market. Zee's share of viewership peaked at 72 per cent in 1995, before competition from other local providers began to eat into its dominance (Page and Crawley 2001: 78). Language wasn't the only reason for Zee's success, however. In Sevanti Ninan's (1994: 161–2) account of the growth of satellite television, an unnamed Zee official attributed the channel's success to the fact that there was no misery on the channel, compared with the developmental images that predominated Doordarshan. Others have noted how Zee packaged itself as an unabashed mass entertainment channel, relying heavily on the popular culture of Bollywood to access the mass market (Page and Crawley 2001: 78).

A Beast called 'Live News' and the Curious Case of the 5-minute Delay

The early histories of Asianet and Zee are representative of the problems Indian broadcasters faced. It was difficult to raise money within India, the global corporations were too powerful, and the state's legal structure had to be circumvented. It was far easier to join either the foreign broadcasters or Doordarshan, and piggyback their coat-tails. NDTV exemplified both these trends. Its evolution clearly illustrates the confusion within the command structure of television that satellite television engendered.

When Doordarshan began its second channel, DD Metro, in 1995, Prannoy and Radhika Roy, who had been producing sponsored programmes on the network since 1988, took a chance and approached Director General Rathikant Basu with a proposal for a daily half-an-hour news bulletin.[18] According to Prannoy Roy, Basu was a 'risk-taker. He could easily have said, forget it, because this is a government domain but he said okay, let's try it.'[19] This was how India's first privately produced daily news bulletin, *News Tonight*, was born. Launched in 1995 on Doordarshan, it was a milestone and its subsequent saga brought together the entire gamut of forces and pressures shaping Indian television at the time. *News Tonight* was given permission as a sponsored programme, which meant that NDTV had to pay a small telecast fee to the network but had a free hand to sell advertising airtime to raise its own revenues.[20] The Roys went to the country's top five business houses with the first year's budget and a simple sales pitch: 'We don't know whether you will gain anything or not, but this is a breakthrough for the Indian media, so support us.' All five companies agreed to share the costs for a year and for the first year, NDTV did not have to bother about incomes and finances.[21]

Getting permission and funding was only the first step and the real challenge came when the programme went on air. The very idea of 'live news' was problematic and got immediate notice from the Prime Minister's Office (PMO). According to Prannoy Roy, the Doordarshan newsroom received an urgent phone call from the PMO on the very first day of the telecast:

> I would say the head of Doordarshan was taking a major risk. A risk to the point where on the first night when I went on air, I said, 'the time is 8 o'clock'—I looked at my watch and said—'we are coming to you live with the news'. Apparently, the Prime Minister's Office people were watching and immediately they started phoning saying, 'Is this live? You can't allow him live'. They didn't understand what live meant. They were just terrified by the thought.
>
> So the next day I managed to do it live again but then I got a call from the head of Doordarshan saying look, I have let you go this far, now don't ruin it all by going live. These guys are terrified about that. Just delay it by about 5 minutes if you can.[22]

Live news was not to be allowed under any circumstances and so after two days of 'live news', NDTV devised a clever stratagem to address the political bureaucracy's fears. This was based on a delayed live telecast, with news bulletins recorded live but broadcast 10 minutes after the recording. The news bulletin would start rolling at 7:50 p.m., instead of its scheduled telecast time of 8 p.m., and be recorded till 8:20 p.m. As the bulletin rolled, it would get recorded on a computer disk, and after a 10-minute delay, the computer would automatically start playing out the recording for telecast. Gradually, the time lag between recording and telecast was reduced from 10 minutes to five:

> So we could technically say that we are not live but nobody could edit it or change anything. It was as live. It was just going, getting cached in the disk and coming out the other end. So one had to go through all these niceties because they didn't really understand

live versus non-live or a five-minute delay or 10-minute delay. What difference did it make?[23]

The system worked beautifully but in 1996, NDTV was kicked off Doordarshan, and an inquiry was launched into its financial relationship with the network.

At around the same time, Rathikant Basu, who as Director General of Doordarshan had commissioned *News Tonight*, was hired as Star's head of operations in India to start a new strategy of localizing content, and it proved a grand opportunity for NDTV. Parallel to the market-driven need for localization, Murdoch had always been fascinated with what he called a 'spiritual calling towards journalism' (Shawcross 1997: 19). The loud political debate about 'cultural pollution' through satellite television ensured that while he wanted to get into news, he knew it was too early for him to get into it himself. A decision was then taken to outsource news, and Rathikant Basu, who had already worked closely with Prannoy Roy, turned to him. Murdoch was to provide the money but, according to Roy, the agreement stipulated that NDTV would have 'total editorial control'.[24]

NDTV, therefore, got a prime-time slot for a daily half-hour news bulletin on two STAR channels—STAR Plus and STAR World. How this arrangement worked in practice tells the story of Indian broadcasting's unique legal structure and the constant hide-and-seek between private broadcasters and elements of the state. To bypass the legal restriction against uplinking broadcast signals from India, NDTV and Star, hand-in-hand with the state, had to resort to a stratagem. When the first daily NDTV news broadcasts started, the bulletin would be recorded in its Delhi studio and the tapes would be rushed to the then government-owned Videsh Sanchar Nigam Limited (VSNL) uplink centre in central Delhi.[25] From here, the tape would be uplinked via satellite to Star's broadcast hub in Hong Kong. Here, the signal would be downlinked, advertisements inserted, and then sent up again for rebroadcast into India. Technically, a government agency was being used to send up satellite broadcast signals from Indian soil, and since these were being sent back into the country from abroad, the law stipulating the government's monopoly over broadcasting any kind of signals from Indian soil was not being violated. The signal was received in India by satellite dishes owned by thousands of cable operators who then redistributed it to individual subscribers.

In principle, this was not a unique practice in satellite broadcasting. Rupert Murdoch had used the same practice in the 1980s with Sky Television, when it was beamed into the United Kingdom (UK) from Luxembourg, to sidestep British cross-media ownership restrictions that prevented common ownership of newspapers and television stations. What was unique in India was the co-option of the state in bypassing the letter of the law.

NDTV was following a model that had already been pioneered by Zee TV when it started daily news bulletins at around the same time as *News Tonight* on Doordarshan.[26] Like Asianet's tapes to Moscow and the Philippines, Zee had found that surmounting the legal guidelines against broadcasting from India had been easy as far as entertainment programming was concerned, but news bulletins could not be recorded days in advance as they needed immediacy. So, Zee initiated the use of VSNL's facilities, a model that NDTV followed when it switched to STAR from Doordarshan.

This system was a fig leaf that barely covered the hollowness of the state's claim to broadcasting monopoly and it was removed only in 1999 when the Ministry of Information and Broadcasting finally allowed private operators to run their own earth stations and uplink from Indian soil.

The next stage in the development of news television came in 1998 when STAR and NDTV decided to start India's first 24-hour news channel to comprehensively cover the 1998 general elections.[27] Star's news channel actually started only as an 'election channel', meant to last for three months to cover the campaign in great detail, but it worked so well that it was never shut down. The success of the election channel resulted in an agreement to continue running it as a 24-hour news channel, STAR News, for another five years. It was an agreement that has been termed a 'one-sided sweetheart deal': STAR was to broadcast, finance, and market the channel, while NDTV was to produce content for a hefty production fee and retain total control over editorial content and copyright.[28]

The round-the-clock channel created a problem for the cosy arrangement with VSNL. The old system of sending tapes was alright if you were doing only a few programmes a day, but it was not geared for a 24-hour channel that never stopped rolling news. NDTV reasoned that the only way around this was to have a microwave link between the NDTV studio and VSNL to transmit programming. Such a licence had never been granted before in India and Prannoy Roy had a lot of explaining to do before suspicious officials. 'Nobody had got a microwave licence at that stage. We convinced them that it's just terrestrial, point to point...from us to VSNL. Instead of us going there and handing over a tape, this is just sending the tape by digital mechanisms. After much convincing they agreed.'[29]

Even then, the old fear of 'live news' remained and this was strictly not allowed. What was allowed was the old NDTV practice of a 5-minute delay in the live telecast. The fear of 'live news' centred on the old phobia of losing control and the reasoning behind it was to allow government censors to sit at the VSNL broadcast centre and make sure that nothing objectionable went on air. The 5-minute delay, theoretically, allowed the censor to monitor every second of every transmission as it went to the satellite, while still having the power to switch the transmission off long before it reached the viewers. For the first five years of its existence, NDTV's headquarters in Delhi had two clocks in most rooms—one on Indian Standard Time, and the other on 'NDTV time', which was always 5 minutes behind. But this was always more of a notional control, than a practical one, as Roy explains:

> Nobody had the time but they wanted control to switch off stuff when they wanted...It was about control...They never did it either because [they were] too inefficient or because who would actually sit there and watch everything going out all the time. And I think by that time we had established a bit of credibility that we are not a wild organisation that says ridiculous things.
>
> But there were times when we went beyond the borders of what they think should not be said...so we would get phone calls but we never changed anything so now they have given up. We got so many phone calls during the Gujarat riots [2002] ...

me personally ... to stop the coverage. We got threats but we never changed anything.[30]

In practice, the censors never sat in, though politicians did try and influence media coverage through subtle and some not-so-subtle means that are common in all democracies.

Much water has flown down the Yamuna since those early years and satellite television has become a powerful factor in the public life of India but much remains to be done. For instance, 15 years after the Supreme Court's historic judgement that freed the airwaves from government monopoly, India is still waiting for a comprehensive new law that covers all aspects of broadcasting, including reasonable content guidelines and cross-media ownership laws. The Ministry of Information and Broadcasting has periodically tried to fill the regulatory vacuum with draft legislation and summary executive directives/notifications. It has often given the appearance of trying to put the genie of broadcasting back into the bottle and the lack of consensus on a new broadcasting bill means that the legal framework underpinning private broadcasting still remains a minefield. For all its flaws, though, the creation of the Indian satellite news industry has been a landmark struggle unparalleled in the history of global news television.

Politics, Democracy, and News

There is much that is wrong with Indian news television. Critics have called it too loud, too sensationalist, too focused on urban middle-class audiences, and overtly obsessed with the holy grail of weekly ratings. All that is true, but a great deal of its problems is linked to a structural problem inherent in the business models of satellite television networks. In economic terms, the Indian satellite networks are inordinately dependent on weekly ratings, more so than their Western counterparts. I have shown elsewhere that this is partly a direct derivative of the peculiar 'illegal' origins of satellite broadcasting in India which meant that channels never had full control of their own distribution and therefore, lost out on a large chunk of subscription revenues. In a market where more than 70 news channels are competing for advertising, the structural economy of television forces many channels to focus on content with the lowest common denominator that will register on television rating panels (Mehta 2008: 140–93). Given the extremely narrow base of these ratings, cricket and Bollywood have emerged as an easy option to register on them. Both have a pan-Indian appeal cutting across socio-economic and regional categories. News of a farmer suicide in Vidarbha may not interest anybody in Kerala, a news producer may reason, but news of the Indian cricket team interests people in every region of India. This is why when national news editors want to lift the ratings of any show, they look towards cricket and Bollywood, and these genres increasingly dominate news space.[31] News, as such, is a commercial product packaged to suit commercial targets.

However, the economic imperatives of creating a market and sustaining it simultaneously drive news channels to create a public sphere, however imperfect. The meaning of the message is not static and takes different forms for different people (Thompson 1995: 34–41). The crucial point is that politics, unlike before, has to unfold in an open arena and in the glare of a new visibility that has a life of its own and is often difficult to control. The media's importance lies not in whether anybody is watching or is getting influenced, but in the assumption of it by political leaders and decision makers (see Schudson 1995: 22–5). It is in this context that television assumes an important role and—regardless of its actual impact on the voting public—becomes central to the political process.

To cite one example of the impact of regional-language networks, when the Bengali news channel, STAR Ananda, started operations in June 2005, it announced its launch by instituting daily live public debates between candidates contesting the Kolkata municipal election. These debates marked an important signpost in the political campaigning culture of the city. They were conducted in the city's open public spaces and took the form of public meetings where sometimes as many as some of 10,000 people turned up as live audiences in addition to regular television viewers. The tapes of that programming make for riveting viewing. They show large public rallies of the kind that are familiar to observers of Indian politics but differ in one crucial aspect: these were joint political events, organized by a television channel, and moderated by a STAR Ananda newsman, as rival candidates debated their political views, while their followers raised lusty slogans.[32]

This was happening in a city which had been ruled by the same political conglomerate, the Left Front, since 1977. The debates unleashed political passions and for the first two weeks, mini-riots broke out during virtually every one of the daily events. Rival political groups attacked each other with swords and sticks. In one instance, petrol bombs were also used. The news anchors were roughed up for daring to ask tough questions, and all this happened on live television. The debates created such a problem that the police commissioner of Kolkata called up the channel and asked it to stop, citing the fear of public rioting. As the founding editor of STAR Ananda, who also anchored these debates, explains: 'It created such a furore and became such an instant hit...I didn't even know...that these two warring groups would come with daggers and bombs, and there was one shoot-out incident...The police commissioner personally requested me 25 times...He said to please withdraw this programme...This is creating hell of a lot of jhamela [problem].'[33]

STAR Ananda responded to the commissioner's suggestion with a public campaign for the strengthening of democratic traditions and debate. The editor went on air with news that the police commissioner wanted the public debates to stop and argued that this was a dangerous precedent for Bengali democracy. The important point here is that this tradition of public television debates was not a Bengali innovation. Hindi news channels like Zee TV and STAR News had run numerous such events in various constituencies during national and state elections across north India in the preceding five years. This is what STAR Ananda emphasized, along with the long Bengali tradition of public culture, adda, and political activism that goes back to the Bengal renaissance of the nineteenth century.[34]

The public appeal to democratic principles and Bengali-ness worked and the political violence ceased within two weeks.

Many localities in Kolkata began to invite the channel to hold similar debates between contesting candidates and that single event turned STAR Ananda into a market leader in the Bengali news sphere. Following this success, a year later, two more Bengali news channels started from Kolkata in 2006 in the run-up to the West Bengal assembly election. These two, Zee's 24 Ghanta and Kolkata TV, both followed similar programming formats of public debate and developed these even further.

Bengali television shows how news television feeds off, and into, liberal democratic values, which themselves are rooted in a long heritage of argumentation and debate. In this context, news channels tap into strong oral traditions and heterodox structures of social communication (Mehta 2008: 230–73) that Amartya Sen has labelled 'the argumentative tradition of India'. For Sen, these traditions are an important support structure for the sustenance of Indian democracy (Sen 2005: 12). Indian television thrives on programming genres that marry older argumentative traditions with new technology and notions of liberal democracy to create new hybrid forms that strengthen democratic culture. Argumentative television fits into broader cultural patterns but the very nature of the medium is such that they mutate into newer forms when mediated by television.

The advent of 24-hour news necessitates a fresh look at what happened to the politics–television equation after the rise of news channels. Twenty-four-hour news introduces the element of permanent publicity and forces politicians to adapt to new forms of electronic mediation. First, 24-hour news makes politicians visible on a daily basis. The kind of high publicity that politicians desire during election campaigns is now thrust upon them on a daily basis. The daily television camera symbolizes the scrutiny of public opinion. Even if that public is a 'phantom' one, the politician has to behave as if it is always there. The demands of 24-hour news force politicians to be on the campaign trail all the time. Anyone who has followed television reporters on their daily rounds of party offices in Delhi knows that it is often the insatiable drive of news channels to 'take the story forward' that induces party spokespersons to 'react' to the latest political controversy. Twenty-four-hour news leads to 24-hour politics.

Political Parties, Regional Languages, and Television

Too much of the scholarship on Indian television has been focused on the national channels. Yet, satellite networks have taken different meanings in different regions and in different languages. By the 2006 Tamil Nadu assembly election, for instance, it had become so important that the Dravida Munnetra Kazhagam (DMK) made the free allotment of colour television sets to every family a key plank of its election manifesto. It is a promise the DMK has begun to fulfil since winning back political power. As Maya Ranganathan (2006: 4949) writes, 'not only had the DMK catapulted "television" into a premier position in the electoral discourse but also granted the status of an essential commodity on par with subsidised rice and reservation in jobs'.

Similarly, Tamil television is very different from television in, say, Chhattisgarh, where broadcast journalists encountered a very peculiar kind of censorship during the run-up to the 2003 state election. Every time any of the news channels broadcast a news item that was even mildly critical of the then Chief Minister, Ajit Jogi, it was blanked from the air. Chhattisgarh viewers watching that particular broadcast would suddenly find their television sets going blank and the pictures would return only 15 minutes or so after the offending news story was over. This unannounced censorship would happen only within the territorial boundaries of Chhattisgarh and television viewers in the rest of India did not encounter this problem at all. This was because supporters of the chief minister had set up a state-wide private television network—Akash (Sky) TV—that bought over, or took control of, cable distribution networks across Chhattisgarh, and this provided an easy mechanism for controlling the broadcast of national news channels within the state's borders. The national networks could be turned off each time their product did not suit the ruling establishment. It was an ingenious form of censorship: it wasn't officially announced, it technically did not come from the state, and there was nothing any of the channels could do about it.[35] The uses, or misuses, of Akash TV became an important part of the Bharatiya Janata Party's (BJP) electoral campaign against Jogi in 2003 and within hours of his losing power on 4 December, its television studios were taken over by a triumphant crowd of the party's supporters.[36] Anecdotes like these reveal the complexity and the centrality of news television across India's regions.

Like the Congress in Chhattisgarh, the ruling Akali Dal has been accused of forcibly capturing cable TV operations in Punjab. In August 2007, the Cable Operators Federation of India complained to the Ministry of Information and Broadcasting of physical threats and arrests of cable operators in the state. Like Jogi in Chhattisgarh, Sukhbir Singh Badal (President, Akali Dal) denied the charges but the parallels were undeniable. Many cable operators in the state were forced to replace the popular Punjab Today channel with the new Akali-friendly channel, PTC, on their prime frequencies soon after the Akali Dal came to power. As one cable operator from Patiala said after being released from prison on charges of violence, 'This is state terror being used against us and the police are being used freely and scores of false cases are being filed' (Chakraborty 2007). Congress Members of Parliament (MPs), now on the back foot, even planned to approach the National Human Rights Commission on the issue, and whichever side one chooses to believe in this dispute, it is undeniable that across India, political parties are taking private television seriously.

What is interesting is that by 2008, numerous channels were openly owned or aligned with political parties. Doordarshan continued to be a state-controlled enterprise. In Tamil Nadu, the DMK has shifted patronage from Sun TV to Kalaignar TV, while the All India Anna Dravida Munnetra Kazhagam (AIADMK) controls Jaya TV. Makkal TV is considered close to the Paattali Makkal Katchi (PMK), while ETV has long had close ties with the Telugu Desam Party (TDP). In Karnataka, Kasturi TV is identified with Janata Dal (Secular) (JD(S)), while the Communist Party of India (Marxist) (CPI(M)) patronizes Kairali TV in Kerala. The Congress has recently backed Jaihind TV in Kerala, while Akash Bangla in West Bengal is controlled by the CPI(M) (Raman 2008).

These networks with political patronage coexist with many other independent ones, all competing for the same markets.

* * *

It is not my claim that satellite television's influence/impact on India has always been positive. Television performs many of its transformations 'subliminally'. Simply by being there, available for viewing, for debate, and for participation, it has effected changes in the way Indians operate in and interact with society. The capitalists who led the move towards private satellite broadcasting in India did not do it for altruistic reasons—their objective was to make money—but their efforts have led to the creation of newer modes of public action and publicness. Television has been adapted by Indian society—by its entrepreneurs, by its producers, and by its consumers—to suit its own needs. Looking to create markets for advertisers, Indian producers and entrepreneurs searched for publics and, as purveyors of identity, they tapped into, but also altered, existing social nodes of identity and communication.

This has not always been rational or 'positive', but it is fundamentally different from the past when television was nothing more than a governmental tool. Television has opened up avenues that previously did not exist and brought many more people into the public arena. This is why Rajdeep Sardesai (2006) has argued that 'The television picture and sound-bite has been one of the most dramatic political developments in the last sixteen years... mutually competitive 24 hour news networks are almost direct participants in public processes: not only do they amplify the news, they also influence it'.

Measuring the political effect of television is, however, an inexact science. Television does not explain every social and political change in contemporary India. To make such a claim would be an overstatement. It is my argument, however, that it is impossible to imagine, or explain, modern India without reference to television, that it just would not make sense without it.

Notes

1. Figures based on 2009 data from Ministry of Information and Broadcasting, available at http://www.mib.nic.in/ShowContent.aspx?uid1= 2&uid2=84&uid3=0&uid4=0&uid5=0&uid6=0&uid7=0 (accessed on 6 December 2009).

2. India had 34,858,000 television sets in 1992 (see Joshi and Trivedi 1994: 16).

3. The National Readership Studies Council (2006: 4) survey estimated a total of 112 million television sets in India. The Indian population in 2006 had gone up to 1.12 billion (Population Reference Bureau 2006: 1).

4. Friedman explains the rise of India and China as major players in the global economy by arguing that digitization and the development of new technology over the past decade has levelled the capitalist playing field so much that it is far cheaper now for smaller local players, and individuals, to compete with bigger corporations. Friedman provocatively argues that this has 'flattened the world', and that big global corporations have recognized and embraced this change in what he calls 'the great sorting out'.

5. Procedural models of democracy focus on the systems and institutions of democracy as symbolized predominantly by the act of voting.

6. At the time, it cost approximately $2 million to lease a transponder for a year and Star accounted for half of AsiaSat's revenues in 1993 (figures are from Ghemawat 2005: 5).

7. Interview with M.S. Kohli, Kohlees DTH Services and Founding President, Delhi Cable Operators Association, in New Delhi, 22 December 2004.

8. Interview with M.S. Kohli, Kohlees DTH Services and Founding President, Delhi Cable Operators Association, in New Delhi, 22 December 2004.

9. No centralized official body keeps a count of the number of cable operators all over India and we have to rely on estimates. The figure quoted here is a reasonable consensus estimate by various television industry analysts, though it came down substantially by late 2004 with the consolidation of the cable industry. According to Star, in late 2004, India still had 20,000 big cable operators and a far larger number of smaller ones. Interview with Peter Mukerjea, Chief Executive, STAR India, 1999–2007, in Mumbai, 12 January 2004.

10. Interview with M.S. Kohli, Kohlees DTH Services and Founding President, Delhi Cable Operators Association, in New Delhi, 22 December 2004.

11. For a detailed analysis of the reforms, see, for instance, Vijay Joshi and I.M.D. Little (1996). Most economists agree that the measures of 1991 were fundamentally different from previous reforms in the 1980s. In contrast, Deepak Nayyar (2006), for instance, has argued that 1991 was not a fundamental turning point, that economic performance had been good from 1980 onwards, and that the failure had been in translating growth into well-being.

12. It should be noted that the annual cost for hiring a transponder came down substantially in later years.

13. Interview with Sashi Kumar, Founder, Asianet, in Chennai, 17 January 2006; emphasis added. Kumar sold his stake in Asianet in 1998 and went on to found the Asian College of Journalism in Chennai in association with *The Hindu*.

14. This was not unique. Jain TV also started its satellite service from Russia on a Russian satellite.

15. Interview with Sashi Kumar, Founder, Asianet, in Chennai, 17 January 2006.

16. Interview with Sashi Kumar, Founder, Asianet, in Chennai, 17 January 2006. The Raheja Group bought a 50 per cent stake in Asianet Satcom and this provided the capital to stabilize the network.

17. Conversation with Sashi Kumar, Founder, Asianet, in Melbourne, 12 December 2005.

18. DD Metro was launched as a satellite channel on 1 April 1993 (Bhatt 1994: 92).

19. Interview with Prannoy Roy, President and Wholetime Director, NDTV, in New Delhi, 28 October 2005.

20. Doordarshan had, and still has, two categories of privately produced programmes. The first is directly commissioned programmes, wherein the network bears the cost of programming and pays the producers a production fee. The second category is sponsored programming, wherein the concerned producers, after getting due permission, pay Doordarshan a small telecast fee and in return can sell advertising for the show to raise revenues.

21. Interview with Prannoy Roy, President and Wholetime Director, NDTV, in New Delhi, 28 October 2005.

22. Interview with Prannoy Roy, President and Wholetime Director, NDTV, in New Delhi, 28 October 2005.

23. Interview with Prannoy Roy, President and Wholetime Director, NDTV, in New Delhi, 28 October 2005.

24. Interview with Prannoy Roy, President and Wholetime Director, NDTV, in New Delhi, 28 October 2005.

25. The VSNL was created as a public sector undertaking in 1986 and succeeded Overseas Communications Unit, a department of the Ministry of Communications. The Government of India first disinvested some stake in 1999 but remained a majority stakeholder till 2002. It is now managed by the Tatas, though the Government of India still has a 27 per cent stake.

26. In contrast to the Sanskritized Hindi of Doordarshan news, early Zee News bulletins were broadcast in 'Hinglish', Hindi interspersed with

English words. Zee claimed this was an attempt to connect to north India's mass audience. News was first broadcast as a single half-an-hour daily bulletin but by 1998, its frequency had increased to five bulletins a day, two of which were in English (personal experience at Zee News, 1997–9). I am grateful to Malay Pradhan, producer at Zee News (1996–2000), for explaining the intricacies in the broadcast arrangements with VSNL.

27. Both NDTV and Zee TV claim to have started India's first round-the-clock news channel. The fact is that NDTV started first with Star News on 22 February 1998. It has also been claimed that BITV's TVI channel which ran from 1996–8, before folding up, was India's first news channel. TVI certainly planned to be a current affairs channel but focused more on programming and never became a news channel of the kind that Star and Zee became.

28. NDTV's production fee was estimated to be roughly $20 million per year (Dalal 2003).

29. Interview with Prannoy Roy, President and Wholetime Director, NDTV, in New Delhi, 28 October 2005.

30. Interview with Prannoy Roy, President and Wholetime Director, NDTV, in New Delhi, 28 October 2005.

31. Interview with Uday Shankar, Chief Executive Officer (CEO) and Editor, Star News, 2003–7, in Shanghai, 22 August 2005.

32. Star Ananda, *Kolkata Municipal Election* tapes, Star Archives. I am grateful to Uday Shankar for giving me copies of these broadcasts.

33. Interview with Suman Chattopadhyay, Founding Editor, Star Ananda, in Kolkata, 22 December 2005.

34. Interview with Suman Chattopadhyay, Founding Editor, Star Ananda, in Kolkata, 22 December 2005.

35. Interview with Sanjeev Singh, Principal Correspondent, STAR News, in New Delhi, 25 January 2004.

36. Interview with Sanjeev Singh, Principal Correspondent, STAR News, in New Delhi, 25 January 2004.

References

Agarwal, Amit. 1994. 'From Rice to Riches', *India Today*, 15 August, p. 76.
Bansal, Shuchi and Brian Carvalho. 1998. 'India's Murdoch', *Business World*, 7 September, pp. 18–25.
Bhatt, S.C. 1994. *Satellite Invasion of India*. New Delhi: Gyan Publishing.
Chakraborty, S. 2007. 'Akali Dal Accused of "State Terror" on Cable Ops in Punjab', 8 August, available at http://www.indiantelevision.com/headlines/y2k7/aug/aug105.php (accessed on 9 August 2007).

Dalal, Sucheta. 2003. 'Is STAR Looking to Replace NDTV...?', 20 March 2003, available at http://www.suchetadalal.com/articles/display/1/282.article (accessed on 20 April 2006).

Friedman, Thomas. 2007. *The World is Flat: The Globalized World in the Twenty-first Century*. Camberwell: Penguin Australia.

Ghemawat, Pankaj. 2005. *STAR TV in 1993*. Harvard Business School Case No. 9-701-012. Boston: Harvard Business School Publishing.

Ghose, Bhaskar. 2005. *Doordarshan Days*. New Delhi: Penguin/Viking.

Jeffrey, Robin. 2003. *India's Newspaper Revolution: Capitalism, Politics and the Indian-language Press*. New Delhi: Oxford University Press.

Joshi S.R. and B. Trivedi. 1994. *Mass Media and Cross-cultural Communication: A Study of Television in India*. Report No. SRG-94-041. Ahmedabad: Development and Educational Communication Unit, Indian Space Research Organization.

Joshi, Vijay. 1994. 'Macroeconomic Policy and Economic Reform in India', *The Journal of the Indian Institute of Bankers-Bombay*, 65(2): 64–72.

Joshi, Vijay and I.M.D. Little. 1996. *India's Economic Reforms 1991–2001*. Oxford: Clarendon.

Manuel, Peter. 1993. *Cassette Culture: Popular Music and Technology in North India*. Chicago: Chicago University Press.

Mehta, Nalin. 2008. *India on Television: How Satellite TV has Changed the Way We Think and Act*. New Delhi: HarperCollins.

Murdoch, Rupert. 1994. 'Power of Technology to Liberate', *The Australian*, 21 November, p. 11.

National Readership Studies Council. 2006. 'NRS 2006—Key Findings', Press Release, National Readership Studies, Mumbai, 29 August.

Nayyar, Deepak. 2006. 'India's Unfinished Journey: Translating Growth into Development', *Modern Asian Studies*, 40(3): 797–832.

Ninan, Sevanti. 1995. *Through the Magic Window: Television and Change in India*. New Delhi: Penguin.

Page, David and William Crawley. 2001. *Satellites over South Asia: Broadcasting, Culture and the Public Interest*. New Delhi: Sage Publications.

Population Reference Bureau (PRB). 2006. *World Population Data Sheet*. Washington, DC: PRB.

PricewaterhouseCoopers and FICCI. 2005. *The Indian Entertainment Industry: An Unfolding Opportunity*. New Delhi: FICCI.

Rajagopal, Arvind. 2001. *Politics after Television: Religious Nationalism and the Reshaping of the Indian Public*. Cambridge: Cambridge University Press.

Raman, Anuradha. 2008. 'Down for the Count', *Outlook*, New Delhi, 12 May, pp. 14–16.

Ranganathan, Maya. 2006. 'Television in Tamil Nadu Politics', *Economic and Political Weekly*, 41(48): 4947–51.

Registrar General of India. 2001. *Projected and Actual Population of India, States and Union Territories, 1991*. New Delhi: Office of the Registrar General.

Roy, Ramashray. 1993. 'India in 1992: Search for Safety', *Asian Survey*, 33(2): 119–28.

Sardesai, Rajdeep. 2006. 'Prime Time Reservation', 29 May, available at http://www.ibnlive.com/blogs/rajdeepsardesai/1/11708/prime-time-reservation.html (accessed on 30 May 2006).

Schudson, Michael. 1995. *The Power of News*. Cambridge, MA: Harvard University Press.

Sen, Amartya. 2002. *Rationality and Freedom*. Cambridge, MA: Belknap Press.

———. 2005. *The Argumentative Indian*. London: Penguin.

Sen, Amartya and Jean Dreze. 1989. *Hunger and Public Action*. Oxford: Clarendon Press.

Shawcross, William, 1997. *Murdoch*. New York: Touchstone Books.

Sinha, Nikhil. 1998. 'Doordarshan, Public Service Broadcasting and the Impact of Globalization: A Short History', in Monroe E. Price and Stefaan G. Verhulst (eds), *Broadcasting Reform in India: Media Law from a Global Perspective*, pp. 22–40. New Delhi: Oxford University Press.

Thompson, John B. 1995. *The Media and Modernity: A Social Theory of the Media*. Cambridge: Polity Press.

Yoshino, Michael Y. and Peter Williamson. 1994. *STAR TV (B)*. Harvard Business School Case No. 394-213. Boston: Harvard Business School Publishing.

8

NDTV 24×7 Remix

Mohammad Afzal Guru Frame by Frame

JOHN HUTNYK[1]

To evaluate the framing of television news in India in the context of developing a theory of media that pays attention to context and form, examples of recent 'terror incidents' and the ways they have been reported, discussed, and presented through television will be considered. This chapter looks, in particular, at issues surrounding the trial of Mohammad Afzal Guru in 2006–7, but also with reference to the terrible Mumbai attacks of November 2008. Whilst tragic in multiple ways, these events are also made spectacular, emotive, and divisive, according to interpretation, by television.

Madhava Prasad (1998: 9) speaks of cinema as 'an institution that is part of the continuing struggles within India over the form of the state', where he identifies a spectrum with 'Hindu nationalism at one end appropriating the fragile national project in an attempt to re-establish political unity on a communal foundation' and at the other end, a globalization that 'seems to be eroding the function of the state as a political restraint on a re-vitalized, rampaging capitalism' (Prasad 1998: 8–9). Participation in the circuits of global commerce, of course, does not come without its ideological inflections too, but there is something to learn from the constant recourse of film studies scholars in India to the state as explanatory framework. Not that other national cinemas cannot be described in terms of a coinciding of cinema and power, as Rajadhyaksha (2009: 11) notes of the Philippines in a footnote and when reporting that Godard could 'startlingly propose in his *Histoire(s) du cinema: Toutes les histories* (1989), that the cinema's birth and death may well be marked as coinciding with the birth and death of the Soviet Union' (Rajadhyaksha 2009: 221). Gore Vidal's (2000) rendering of Hollywood might best exemplify the same argument for America, and Rajadhyaksha nominates Dave Morley's (1980) *The Nationwide Audience* as the key British text, but it is in Indian film studies that habitual recourse to the state story seems most evident. What can be said here of cinema also holds for television; though remembering that cinema and television is not the same as television news, I still take the suggestion of Raminder Kaur and Ajay Sinha (2005: 23) seriously when they offer an analytical perspective that notes 'the interdynamic relationship between the local and the global, the national and the international...to draw attention to the audio-visual and cultural economies...and the flow of representational capital and technologies of production'.

24-hour Theory

I do not think media theory alone is adequate as a theory of the media, so I will rely upon the work of Prasad, Rajadhyaksha, and Kaur and Sinha, alongside people like Gayatri Chakravorty Spivak (1999, 2008) and Rustom Bharucha (1998), as well as citing theories of 'attention', such as Bernard Stiegler (2009[1996]) and that of Jonathan Beller (2006). These authors are influential, I contend, because their work is much more than media theory, yet media are necessarily their stock in trade. Spivak (2000: 307), for example, writing of 'Indian Modernity' as 'represented in videographic news', mentions Kashmir and the film *Roja* (Mani Ratnam, 1992) as 'contextualized by the fierce near-Fascist nationalism daily shown on Indian national television'. Bharucha (1998: 115), also mentioning *Roja* and Kashmir, says we 'need to confront that dangerous border where nationalism becomes fascism by questioning our own complicities in the legitimization of violence around us'. I choose these two mentions of Kashmir because the example I want to take up—a series of violent incidents and events presented to us on television—has its origins in part in the Kashmir question, but also because a scrupulous critical commentary on Indian modernity relies upon the kind of contextualization these authors provide. I will not, however, have much to say of Kashmir directly. It is more than 20 years since I visited the region and militarization and the crisis of curfews, confrontation, and bombing has worked to transmute the tourist resort into a vale of tears. I think one of the complicities we need to attend to is that Kashmir is nowadays a code word for many people, much more than it is a place. Kashmir, largely through the news, has become a cipher for something else, a frame for discussion.

Learning from Bharucha and Spivak, an approach to the media that attends to presentation, to framing, to performance, and the way the news is presented as news, also deserves attention. What first strikes me as apparent, but often necessarily ignored, in media presentation is to look closely at just what is presented on the screen. One way to pursue media analysis is to examine station identification, presentation formats, props, and styles of news for clues to what sort of media phenomenon we are examining. The obvious things to look for here are the slogans and catch phrases of media news. Most revealing of what I mean here is the possibility of an analysis of station 'idents' and slogans such as that of NDTV's strapline, 'NDTV 24×7—Experience. Truth First'. It is possible to raise a number of questions here to do with the origins of the name NDTV. Derived from what was initially a content provider company for Doordarshan in the days of state monopoly, New Delhi Television (NDTV) was a private concern run by Prannoy and Radhika Roy, initially broadcasting a half-hour news programme, called *The World This Week*, from November 1988 until the mid-1990s. From *The World This Week* to 24×7 is perhaps not so huge a temporal shift, but it is possible to say that the Roys brought with them a considerable track record, if not a '24×7 Experience'.

In terms of a strict reading of time, one experience that was carried from the weekly news segment on Doordarshan to the 24-hour television format version involves a quite curious delay. Television scholar and journalist, Nalin Mehta (2008: 76), identifies NDTV as the best 'place to begin the story of what was happening to television news within the larger framework of television expansion' because of something the channel introduced that greatly troubled the Prime Minister's Office (PMO, Rajiv Gandhi)—the idea of 'live news'. On the first day of the telecast, as Prannoy Roy describes it to Nalin Mehta, the announcement that the news was 'coming to you live' caused consternation in the prime minister's office, prompting anxious calls from the 'terrified' PM's aides (Mehta 2008: 77)

NDTV's solution was to delay the live telecast by 10 minutes, broadcasting at 8 p.m. a programme recorded at 7.50 p.m. The story of the multiplication of television channels through satellite in India is already well known (see Gupta 1998; Mankekar 1999; Rajagopal 2001), but it is curious that the delay in live transmission of the news was still in place when NDTV moved from Doordarshan to Rupert Murdoch's STAR TV network to start the first 24-hour dedicated news channel (according to Mehta [2008: 81], this was to cover the 1998 general election—though the claim to be first is disputed by Zee, see Mehta [2008: 329]). The delay was slightly shorter on NDTV's work for STAR, but all NDTV offices had two clocks in each room, according to Roy, 'one on Indian Standard Time, and the other on "NDTV time", which was always five minutes behind' (Mehta 2008: 82). Roy insists that despite calls—for example, during the Gujarat riots of 2002, including threats—the station did not stop its 'live' broadcasts. That is, live for those in the room on NDTV time; for everyone watching it, it was five minutes later. Consider again the NDTV slogan, 'Experience. Truth First', and look at where the full stop has been placed, directly after experience. You will not, as a viewer, experience truth first, but rather the experienced news editor looks at the truth first and then broadcasts.

What is screened shapes understanding. I think it significant that NDTV 24×7 shifts from a reporting-as-public-service function in the early days, to something that updates what some might call national project television (Doordarshan) with a much more commercial attention–economy focus. Speaking of NDTV's weekly debate format show, *The Big Fight*—to be discussed further later—Mehta (2008: 255) says that television 'turns politics into spectacle, but politics has always been about spectacle'. He interviews NDTV Managing Editor (1997–2004), Rajdeep Sardesai, who says:

> TV is now increasingly entertainment. News is entertainment. You have to create some element of entertainment...people shouting at each other...or some kind of conflict. It is not always about information. I am not saying in the *Big Fight* you don't try to inform but if the entertainment element was not there the programme would probably not have survived. You have to package it...First Punch, Second Punch...Otherwise who will see? There has to be some heat. (Mehta 2008: 255)

Even if NDTV is not watched by everyone all the time (not everyone is entertained by this format), it is still possible to elaborate what is presented on the channel as indicative of a certain interpretive agenda without falling for the rhetorical justifications of management. The suggestion that the news be entertaining is all well and good, but the format of debate itself is not transparent. Just as the name NDTV makes visible but does not discuss the 'New Delhi' view of India—where New Delhi here could be a code for a politically centralist, nationalist, and parliamentary project—so too the 'First Punch, Second Punch' format of the debating chamber does not draw attention to the 'entertainmentization' of contested information. To do so would, of course, undermine any pretence to newsworthiness or relevance, which would be another kind of television. Other kinds of television also participate in the ideological formation of the national and the international—game and reality TV shows just as much as movies, all have been often examined in this way. But reality TV is not (always) as horrific as the news. To turn on the television and see that it is always on, that the news never sleeps, and that the most monstrous atrocities, crimes, injuries, deceits, and iniquities are played over and over as current affairs, this is the most grotesque consequence of news as entertainment. Our concern does not need to be only about the degree to which we are appalled or inured to terror attacks, but also about how this entails acquiescence to the routine hype that promotes the state security regime which provokes such attacks, the accepted surveillance and the constraints on civil liberties that deserve our contempt but are reported as initiatives of government, the pathetic and transparent lies of those who operate detention centres, extrajudicial assassinations, special renditions, black ops, and in fact, the whole militarized counter-intelligence terror regime that is the staple of daily news in uncertain geopolitical times. That's entertainment.

Trial by Media

On 13 December 2001, a little over two months from another now overdetermined date of significance in New York, five men (at least) piled out of a white ambassador car that had driven into the grounds of the Parliament building in New Delhi. The winter session was on, and, as if out of the latest blockbuster movie, these miscreants, with guns blazing, attacked and killed nine people, and then themselves died in a hail of bullets, failing to set off their car bomb because they had damaged their detonator in a collision with the President's parked vehicle. As a consequence, the military was immediately deployed and the border with Pakistan sealed, terror legislation was enacted and terror threat level escalated for a year, along with a high-profile court case and debate all through the press.

Alleged accomplices of the attackers were subsequently arrested. There were statements from the inspector general of police declaring that all hands must initiate a national effort, the nation appreciates the sacrifice of the police who left no stone unturned, etc. Many commentators have said that 13 December was a fairly incompetent raid, and the news channels reported it as such. The accomplices were presented as dupes or clichéd troublemakers, no match for the intrepid security forces—as I said, these were 'miscreants', also 'terrorists'. However, some among the commentators, Arundhati Roy, for example, in *The December 13*

Reader (2006a), have questioned the swift 'case cracked' response of the police in arresting and bringing to trial the four accomplices. The *Reader* serves to expose contradictions and inconsistencies in the case, as does *Manufacturing Terrorism: Kashmiri Encounters with Media and the Law* (2006), a commentary on Kashmir by Syed Bismillah Geelani, columnist brother of one of the four accused—Syed Abdul Rahman Geelani (who was a PhD student at Delhi University when arrested on 14 December 2001). These publications raise a whole series of disturbing questions within a wide public debate where the 'facts of the case' have become fairly common knowledge, but also have drowned somewhat in a news media circus. Of significance for the case I am making, as something that reveals the spectacular economy of newscasting in this scene, on NDTV, Vikram Chandra (not the novelist) hosted a 'Big Fight' teleconference in a boxing ring setup. This big-money telecast sporting metaphor surely illustrates the stakes involved.

Three of the cases, including Geelani's, were eventually dismissed, and only Mohammad Afzal Guru was found guilty and sentenced to hang, so as to appease what the presiding judge would call the 'collective conscience' of the nation (Supreme Court Judgement, 4 August 2005).[2] Mohammad Afzal, also known as Afzal Guru, had been a 20 year old border crossing militant youth in Kashmir, but had 'surrendered' to authorities in the early 1990s and then enrolled at university in Delhi. His experience with said authorities was, of course, not all pleasant as he was first tortured in the mid-1990s, and found to be 'clean' by one Davinder Singh. It was Singh who proudly announced in a television interview that he tortured 'for the nation' (as cited by Roy, 2006b). Disturbingly, the method of 'chilli and petrol enema' was the 'cleansing' facilitator of confession (even petrol seems to get the red hot vindaloo treatment) when Afzal was picked up and interrogated after the 13 December 2001 incident. Afzal's video 'confession' implicating himself in the raid was later judged by the Supreme Court to have been illegally obtained, leaving his conviction to be based upon serious, yet circumstantial, factors. The case against him now revolved around three pieces of evidence: a shopkeeper seeing Afzal buying the mobile phones and explosives used in the

Parliament raid (the phones left in the Ambassador car); his renting rooms to the five men (or were there, in fact, perhaps six attackers, as closed-circuit television [CCTV] footage shown immediately after the attack seemed to suggest, though this footage was later restricted); and Afzal having possession of the computer upon which the fake ID cards were made.

Found guilty and slated for hanging on 20 October 2006, NDTV 24×7 screened Afzal's video confession. They did so without mentioning that this was, by then, a five-year-old and discredited piece of footage. After the screening, it emerged, from reports by a police inspector, that this video was version three of a much-rehearsed 'statement'. NDTV omitted mentioning the Supreme Court's rejection of the footage, all the while allowing an on-screen short messaging services (SMS) commentary to announce the 'collective conscience' viewpoint: that terrorists should hang, that Pakistan was behind it all, that the national institutions of law must be respected, and due process must take its course. At this time, Afzal's execution was being reviewed on appeal, but the SMS poll seems to have decided his fate. The death sentence was upheld by the Supreme Court on 12 January 2007. Only a plea for clemency by his wife forestalled the hanging. Two sitting presidents, former President A.P.J. Abdul Kalam and former President Pratibha Devisingh Patil, who are the responsible figures with regard to the final decision on hanging, could not make a decision.[3]

I am particularly interested here in the justice process as it was played out through the televisual public sphere. Arundhati Roy and other prominent intellectuals spoke out, and television stations set up opposing views and spokespersons of note. This was an elite mohalla discussion with commentary by SMS and phone of the lynch-mob variety. *The Big Fight*, hosted by Chandra, claims to 'pit those on opposite sides of an issue to against each other' in a 'thorough 360° view of the key national or global issue at hand'. This 360 degree theatre can be hilariously and disturbingly literal, such as the discussion of the case of Afzal. Certainly not something to be trivialized as sport as it has to do with a man's life and a nation's attitude to the death penalty, this *Big Fight* boxing ring performance reduced issues of crucial significance to questions of ratings and used a format that replicates celebrity games or quiz competitions.

The SMS participation as the semblance of reasoned views (adding a necessary frisson of controversy) also seems quite problematic. Mehta (2008: 6–8) notes the thousands of SMS messages solicited by the stations as 'television actively sought to construct a united nation' by way of this 'new and revolutionary theatre' in India. Our *Big Fight* host, Chandra, has been uniting India through NDTV since 1994 and has been its special correspondent, covering the Siachen and Kargil wars, and the conflict in Kashmir. He won the Indian Television Academy Award 2008 for 'Best Anchor for a Talk Show'—award shows themselves being part of the ideological manufacture of 'Experience. Truth First'. Of course, SMS should be considered a part of participatory media which requires that we should also evaluate the framing of 'vigilante reporters' armed with mobile phone cameras contributing to what must now be the world's largest 'tele-democracy' (Mehta 2008: 257). Television worldwide also provides, and may be examined in terms of, an emergent public interface of which SMS is only a part (also multi-screen, phone in, video contributions). This is, in fact, an old model, largely pioneered by music television stations and the shopping channel, only recently adapted to news, and especially in India. The commercial and pop origins of this managed interactivity are significant but should not be overplayed. There has been considerable complaint in the past that television was a 'push' media. The 'pull' factors discussed here, however, are significantly constrained by form, so that it remains an open question as to whether these technologies imply a transformation of media space or otherwise. Surely, we are not able to judge if the popularity of the NDTV style of current affairs debate should be attributed to an 'argumentative tradition' or to versions of '*adda*' as suggested by Mehta (2008: 245), or if, as seems more likely on the other hand, the incorporation of SMS and commercial consumer consolidation mixes with national security and anxiety of the incorporated public to provide a mere semblance of debate in the 'Shining' India of 24-hour news.

As time passed, Mohammad Afzal Guru continued to cool his heels 24×7 on the death row, and experienced this as a kind of torture. He went so far as to request the former President, Pratibha Patil, to make her decision. The trouble was that there were 50 death penalty clemency applications in the queue:

> Parliament attack convict Afzal Guru is not the only man waiting for his death pardon to be reversed by the President. Sentenced to death in 2005, Afzal Guru may be the face of a raging debate on death penalty and clemency. But a Right to Information activist who petitioned the President's office found that the number of pending mercy petitions of those who want their death sentence changed to life is as many as 50. (NDTV, 27 August 2008)[4]

In a perverse mode, I have imagined how television might handle the eventual hanging of Afzal. I have suggested there be commissioned a satellite hanging channel slot that could aggregate scenes such as Saddam Hussain's hanging, movies like *Dead Man Walking*, the story of Bhubhaneswari—this is a reference to Spivak's 'Can the Subaltern Speak?' essay and its rewrite in her book, *Critique of Postcolonial Reason*—the story of a woman who killed herself in order not to betray her revolutionary comrades but was still misspoken for by family and history (see Spivak 1999: 306–8) and a 'reality TV' scenario of the Afzal appeal, again with SMS voting to allow the people to decide. I have suggested this, of course, with considerable bleak irony and do not hold out much prospect for the various anti-capital punishment campaigns that periodically arise since their media face does not chime with national and international requirements.

Instead, NDTV goes on to host the successful show, *Airtel Scholar Hunt: Destination UK*—a mobile phone company-sponsored reality TV vehicle to bring a media and cultural studies scholarship student to Cardiff and management students to Warwick, etc. A great publicity coup for the United Kingdom (UK) teaching factories, in which the cultural construction of fantasy India, a UK vice-chancellor's dream of subcontinental expansion (of the teaching factory), and a tamed public sphere without a hint of critique proceeds apace (it was once thought the university was a place for rampant intelligence, now its sold like soap on television, not even as smart as a quiz show like *Crorepati*). It has not gone unnoticed that television has been particularly bland in its programming in the wake of the invasions of Afghanistan and Iraq. For example, Daya Thussu (2007: 104) quips that 'Even serious news networks, such as NDTV 24×7, have increased their quota of talk and chat shows'. There has been, of course, significant media coverage of the hanging of Saddam, as well as considerable mention of the fiascos of Guantanamo and Abu Ghraib (and Sangatte, Harmondsworth, Woomera, special rendition, WikiLeaks, etc.), but predominantly, television tends more and more to avoid even the talk show news format. Accordingly, it is not accidental that we see a saturation of domestic programming with game shows, reality TV, police dramas, and parodies of the format of the court highlighting questions of judgement and voting. This is what Avital Ronell (2010: 52) diagnoses as a 'repetition compulsion' that is an 'incessant petition to the law' which is, in turn, 'frustrated by the endless postponement of justice'.

As a coda for this section, think also now of how this played out, yet again, in relation to Mumbai in 2008. Another NDTV *Big Fight* debate on the terror attacks in Mumbai lined up a number of prominent pundits in similarly tactless, though perhaps more atmospheric, surroundings. The pundits—including Najma Heptullah, Imtiaz Ahmad, Aamir Raza Husain, Waheeda Rehman—sat in the open-air garden outside an impressively lit Indo-Gothic building, possibly a hotel forecourt. There was a 'studio' audience in attendance, though they were not inside the studio, and the compere chaired a passionate, wide-ranging debate that only sometimes steered beyond—though, importantly, it did go beyond—the protocols of expectation: speakers questioned the requirement for all Muslims to show remorse that the attackers were Muslims; the requirement for Muslims in India to show that they distance themselves from Pakistan; routine denunciations of the state of Pakistan; praise for the Indian police and anti-terror forces; concerns about lapses in security (solution, better funding of security forces and more training); calls, from the compere, for the guests to think of 'ways we can channel our anger' and be 'united as a nation'; and so on. Watching NDTV on another occasion offered further opportunity to see the nation affirmed and confirmed through the response to terror—this time viewers were able to watch *Walk the Talk* with J.K. Dutt, now retired Director

General [DG] of Security, speaking of Operation Black Tornado, the response by the security forces to the attacks in Mumbai in November 2008 (NDTV, 4 April 2009),[5] once again the SMS facility, the backdrop, the polite reporter allowing the DG to 'finally' speak out, always in a tone that affirms the 'job well done' congratulatory and civil society affirming success of it all.

SMS Bomb

Our complicity in 'the legitimization of violence around us' (Bharucha 1998: 115) is something achieved by way of our television screens. We are the audience that watches and by watching, convenes the community of television within which these complicities are played out, and indeed enact the violences they depict. It is clear by now that the varieties of the 'terrorist' we know so well are, in fact, shaped—constituted—alongside the equally fictive moderate citizen, or television audience, that we are presumed to be. The terrorist, of course, has bombs strapped under the shalwar-kameez, and the other is in constant SMS contact. The audience, both nationalist and economically aspirational–ascendant, are conjured into existence by way of engagement out of proportion to the extent of actual terrorism (and thus tempting fate, manufacturing expectations). Framed above by the logo of NDTV and captioned below by a rolling ticker giving headlines and stock prices (though too fast to give any meaningful correlation between shares and events), the low-resolution CCTV footage of the attack on the Parliament is contained in the same frame as the honoured guests on *The Big Fight*. Equivalences are not intended, but inevitably made—viewers become inured by way of format. Indeed, viewers are encouraged to be vigilant and provide commentary, perhaps on occasion even be the eyes and ears of NDTV, recording mobile phone video of events and providing it to air—the development of the 'citizen journalist' and 'vigilante reporter', a previously rare occurrence, nevertheless now anticipated as the way of the future and with some fanfare. Terror events will catch us all, live reports by way of networked personal media produced by and available to viewers who could just as easily be caught up in events as be watching at home. In this way, Indian modernity and postmodernity are both manifest in a low-level everyday fear or anxiety about security and prosperity. This security and prosperity, both rely upon electronics, and thus a convergence entails the simultaneous promotion of robust surveillance and a widespread consumer sector. The work of promoting this doubled platform combines design, framing, training, organization, intellectual labour and celebrity punditry, infrastructure and planning, coordination, and luck—none of which appears as a seamless whole, but which nonetheless combines to great effect.

Terror is presented as effecting everyone, yet it comes from outside—it is both familiar and alien. It is to be discussed, analysed, and detailed. There is a concerted effort to combat and contain it—inspectors of police are interviewed, experts arrayed and displayed, pundits confer with their close-up cameras. The correspondence of despair and celebration is written in the meta-text of NDTV. The double façade of presentation on screen—the double framing of each item, not only the content of what is said—is the preserve of media, and especially news, when it stages traumatic events as shared, and reassures us that everything will be OK, that the nation is defended. Terror strikes, police cordons go up, bombs fall, politicians debate. Stiegler (2009[1996]:115–16) tells us that the media 'co-produce' what happens. Inside this theatre, there is no place for recognition that this framing is constitutive of terror; that the police, the promoters, the formatting specialists, the political grandstanding, the policy platforms, the talk back style, all this legislates terror and bombing. The decoration of the screen with shocking, gruesome horror—news flash, breaking news, scoop—manages this uncertainty with the live ticker, updated scripts, and item 'idents' that reassure continuity, even while leaving ambiguous gaps in the record. Our co-production as viewers is also involved: we contribute attention by looking (and also, perhaps more problematically, we agree to a kind of inattention), internalizing the parameters of 'debate' as if The Big Fight format were our own, as if this were the nation (or the planet) as we would have it, as it would be, never otherwise, just so. It is the obligation of critics to open up these problems, not to let the frame close them down.

NDTV is television in the service of selling mobile phones and demonstrating their functions to the enormous consumer classes (the Airtel and SMS links). That this is then grafted onto the current incarnation of the 'nation' or national–modern as subject only updates Nehru with Noida. The nation-building project becomes the national business model. This, however, is not merely a national question. The diaspora is accessed now as NDTV 24×7 becomes available in the United States (US) and Europe.[6] Again, the format is 24-hour news, with an added strand of reporting from home to those non-resident Indians (NRIs) abroad or who have business interests in the subcontinent, and the occasional media-interested cultural studies scholar. The format of *The Big Fight* debate remains popular internationally also, perhaps because, like sport, the action is always at the spot scheduled and covered by the cameras placed on location in advance.

There are protests of course, duly reported. Dissident voices are part of the entertainment of *The Big Fight*. There are exposures of corruption that lead to resignations—every week a new scandal—or apologies. Government leaders may be voted out of office every four or five years—only to be replaced by another horror show compere. There are theorists of change, and sometimes even the advocates of revolution may be invited on stage to be interviewed—for example, French philosopher, Alain Badiou,[7] as well as then Nepalese's Maoist Prime Minister, Comrade Prachanda, have appeared on the BBC, and Prachanda was seen on NDTV reports where he was said to 'clearly mean business' on a visit to recruit Indian investment.[8] As ever, the format of the report and the time slot—in Prachanda's case, a breakfast news item of 45 seconds—determines the extent of the discussion. It is a matter of 'First Punch, Second Punch' and then, there is a restoration and return to the normative format.

Fascist TV?

Where have we seen this before? This horrible combination of police violence, constant surveillance, and bureaucratic proceduralism—sensation and formality, the script of all news programmes—is nothing if not a latter-day Gestapo operation.

Channeling Cultures

Bharucha (1998: 116) cautions that 'The charge of fascism…can be a violence in its own right, and therefore the word should be used sparingly and consciously. While acknowledging the burden of the terrifying legacy, one should not censor it automatically from contemporary usage.' In this context, in his book on social activism in India, *In the Name of the Secular*, in a particular nuanced passage in the essay, 'On the Border of Fascism: The Manufacture of Consent in *Roja*', Bharucha (1998: 115) cites Chomsky and writes that: 'nationalism is mediated and disguised through layers of cultural expression, which have been consolidated through a "manufacture of consent" engineered by the local agencies of the State in the market and the media'.

If we must be careful not to make the charge that there are fascists in the sense that there is a brown (or saffron)-shirted phalanx marching towards a pogrom, we can certainly speak of fascist structure to a system that has rewired social life in the manner of the work camp and the concentration camp and how—this is the most grotesque element—we have become more and more acquiescent towards the impossible outrages that are telecast before our eyes. We go on working and concentrating while persecution is made routine, in our name, in the name of the nation, the public, security, or peace. The brother of one of the accused accomplices of the Parliament attackers of 13 December 2001, Syed Bismillah Geelani, writes that Kashmiri Muslims are often portrayed in the media as terrorists (Geelani 2006). 'Films like *Roja*, *Mission Kashmir*, *Maa Tujhe Salaam*, *The Hero* and even TV serials have systematically constructed this image'. He reports that in a 2005 serial on Zee TV, called *Time Bomb 9/11*, Osama bin Laden himself surfaces in Kashmir (Geelani 2006: 26). The all too common police procedures of torture, interrogation, and detention, described by Geelani, are harrowing at the same time as they are just what we have come to expect. There are also a number of extrajudicial killings by police of Kashmiris in 'encounters' (Geelani 2006: 94) of the sort made famous in an earlier era when it was Naxals that would meet such fate at the hands of the state, duly reported as another triumph for order. Such horrors—the bombings, detentions, imperial wars, fratricidal aggressions—do not disrupt the rampant pursuit of wealth and the subservience of political figures, as executive committee of the bourgeois class, to nothing but the facilitation of that wealth. Even communist parties in Bengal encourage big capital today, and the slaughter in Nandigram is the consequence.[9] The 24×7 talk show is the bureaucratic form of the parliamentary fiasco which provides the unedifying spectacle of a bland and phantasmatic version of politics alongside entertainment—as containment of debate. A 'big fight'.

But too easily, the talk show abdicates the responsibility of the critic to tread carefully, thoughtfully, yet forthrightly into the political scenarios that critical thinking must address. Bharucha (1998: 133) points out the dangers of the 'seemingly reasonable and self-righteous' stance of the commentator who end ups contributing to 'thought control' and the manufacture of consent (*pace* Chomsky) by playing within the 'rules of the game'. Polite discussion and conversation can be applauded and declared necessary, and I do not always want to disagree, but too much well-tempered conversation could also leave everything intact. In such circumstances, we remain unable to address necessary redistributions (of power, of economics, of desires), we remain uncomprehending of what would be adequate to help win through to a radically different world, and in the end, we serve up only the oh-so-satisfying prospect of our being knowingly in the right, while all around innocents are hanged. The really existing right as such is in ascendancy (as in the racist parties elected to government, the Bharatiya Janata Party [BJP] in India, the rise of the British Nationalist Party and the English Defence League in the UK, and the ultra-right in Europe). Bharucha (1998: 133) uses in expansive but judicious tone when he says: 'Obviously, there are different levels of expertise in reinforcing the premises of the State, but they meet through a common ground of complicities that are manufactured not by force, but through the most open and, at times, inane of exchanges.'

The inanity Bharucha has in mind here is that of the 'magazine culture' which propagates vacillation as a norm. This has been often mentioned, for example, by Thussu (2007: 5) in his study of satellite television where television news seemed 'to take on the worst aspects of the tabloid newspapers' and became 'a powerful discourse of diversion...taking attention away from, and displacing from the airwaves, such grim realities [as] neo-liberal imperialism...tyranny of technology...freemarket capitalism and...consumerism' (Thussu 2007: 9). In my first year of university, in a journalism class that I ultimately failed (taking anthropology as a catch up), I was taught that the formula of good writing was to argue both sides without bias so that a reader with a reading level of 14 years could comprehend the debate. Even this could be a high standard for some parts of the media today.

Bharucha (1998: 134) demands a writing that is not subsumed by the 'trivializers of free speech in the mass media' and thinks that something more can be done with other forums, including 'songs, poems, pamphlets, documentaries [and better films than *Roja*]'. These might provide a more meaningful and detailed picture upon which something 'beyond crude associations with terrorism' would be demanded and acted upon by 'all citizen groups and public forums in India [and elsewhere] concerned with democracy'. In this comment, Bharucha is talking particularly of Kashmir, but with wider relevance, his assessment that it 'is shocking to realize not only how little we know, but how we *accept* this fact' (Bharucha 1998: 134) might be considered equally important in many other theatres. We know we are ignorant of the way the war on terror has transformed our lives, but with our heads down, we tend to ignore it.

Is it that the images and the effort have become incomprehensible, or that the atrocities have been forcibly banalized? The endless wash of images no longer shock, the everydayness of camera phone footage, the seamless 'news' without commentary (but plenty of diagrams), in the press or television. In Hanif Kureishi's (2008) novel, the main character (and psychoanalyst–murderer), Jamal, has a friend called Henry who wants to go to Kings Cross to lay flowers after the 7 July 2005 London bombings. He says:

How can I stop thinking about the horror of those bomb-blasted trains, the ruined bodies, the cries and moans and screams, which segue, in *my* head at least, into the diabolical killings of civilians in Baghdad—severed heads, blood underfoot, children eviscerated, limbs blown into treas. Could only Goya grasp it? Why are we making this happen? (Kureishi 2008: 314)

That he can ask the question—an intellectual playwright—is indicative of its own malaise. The conversation does not broach an alternative even when critical.

Attention!

On NDTV, nothing unusual can happen. The debate is already scripted. The separation between extreme entertainment and considered argument has been fudged by lazy and cynical media operatives, and the national agenda that is, however, not an agenda worked out with a 5-minute delay by the prime minister, but a national project, of 'Unite the Nation', that deploys anxiety and hatred—of Afzal, of calls to 'hang him'—links up with a global media thriving on the 'same' anxieties, the same 'Unity'. All the better then that premium commercial subscription rates are achieved through sensation–attention, and the more controversial the debate, the greater attention from a wide and diverse audience and the greater containment (this is, not to see this audience as conflicted and confused, but the format possibly encouraging traffic in, not critique of, the format). Here, some say attention is the premium; I would add that there is also a production of an attentive inattention. Something like this is suggested by Beller (2006: 28) in *The Cinematic Mode of Production*, where he argues for the 'productive value of human attention' and says that the labour theory of value must 'account for the systematic alienation of the labour of looking' (Beller 2006: 23). He suggests:

> equally significant [is that] in viewing the image, we simultaneously and micrologically modify ourselves in relation to the image as we 'consume' it—a misnomer if ever there was one, since images equally, or almost entirely, consume us...this production of both value and self (as worker, as consumer, as fecund perceiver) through looking...[means] visual culture must be set in relation to the development and intensification of commodity fetishism. (Beller 2006: 24)

Beller's work accesses another model that can supplement media theory for a theory of the media. The media work offered by Marxism is of course vast, and a survey of its parameters is well beyond the scope of this chapter (see Wayne 2003; for India, see Prasad [1998] on subsumption, among many others), but relevant to this commentary, there are some staple perspectives offered by Marxist film studies, redirected to television as I have hinted in cursory form earlier. More explicitly, the task set out is to take the appearance of news and examine what kinds of labour go into its production. There are new elements, and old. I think it is the case that a series of sensational 'breaking news' terror events have transformed the appearance of news, and yet, often, nothing much happens on screen—often the appearance of news is still, distanced, nothing to 'eyewitness' or 'experience'. Considering the few occasions where mobile phone uploads of CCTV footage are offered, usually the 'same' images over and over on very high rotation, the 'live' aspect of the news story consists of a presenter standing before the police lines, or the cordon tape, reporting to camera. The camera will very often then zoom in over the presenter's shoulder into a cleared cordon space while we are told what we are seeing—a distant building with smoke, a ruined bus, the smoking towers. A city where nothing happens, live.

Of course, uploads must be uploaded, and framed. They must be introduced, they are captioned. Some presenter—formerly a journalist, now a desk-bound reader of the tele-prompt—is required to decipher, and speak over, the images on the loop or the live scene shots of the requisite police stand-off, ongoing investigation, or security barrier. Such imageries are of course 'news', but someone is editing them, framing them, adding a station identifier, and an item logo is superimposed, and elsewhere the ticker strap info must be typed in, the stock prices checked, the wiring to the stock exchange maintained. Cameras must be set up, purchased, built, tapes stored, catalogued, reviewed. And satellite uplinks are not spontaneous, there are shifts of workers (it is 24×7 recall) producing the news that the audience attention then consumes—even where the audience is solicited to participate and gives their attention for minimal subscription charge, this labour is not offered free and is certainly not the supersession of paid skilled work of many kinds. It is true that increasingly, the formatting, idents, studio props, and so on are always prepared in advance, as are the trucks, catering, tailoring (for the presenters' suits), and the research that identifies which pundits might be called upon from the little black book of punditry, but even all these events are produced. Nothing is happening on screen—the camera stands still at the side of the police chief managing the barricade, images of the burning hotel, the damaged bus, the shooters, the towers, all these are what Stiegler (2009) calls a 'tertiary retention'—but this does not mean the economy of contribution has changed the ideological field.

Attention to the attention mode of production would not displace or replace older prevailing productive regimes or critique (of manufacturing, ideology, nation, service sector, digital) but offers a frame to think the real subsumption of more and more parts of life to a commodity logic. All the time we are tuning in, we are further attuned for training for a constant contribution to consumption production—and even as we participate in complicit unacknowledgement in yet another round of surplus extraction and valorization, the attention we reflect back upon television only confirms this exemplary form. Television is mundane and emblematic of our circumstances, even as the debates are staged.

It is still the case that in the mix of multiple satellites and constant 'debate', the evidence before our eyes ('Experience. Truth First') means that all the work that goes into representation of these tragedies can be understood as necessarily contained by a modernity that presents itself, for a certain constituency (of Indian television viewers and internationally), as the work of the nation. What it also, more worryingly, means is that in screening and containing anxiety in these formats, the possibility of other Indias or of an India not the same as the one scripted in the national imaginary, is left off screen. Most relevant here are Spivak's (2008) commentaries in *Other Asias*, and especially her consideration of which parts of Asia register *as* Asia, and the chapter, 'How to be a Continentalist', where she argues—in telegraphed form: 'There is no original unity to the name "Asia". When we claim the name today we are divisive. To repair this, Social Sciences and Humanities must come together. The production of knowledge must be supplemented by the training of the imagination' (Spivak 2008: 209). So too for other alternative worlds, other possible lives. To combat (and repair) this, a concerted effort to tamper with the framing of terror might do three things: first, develop a global postcolonial studies that can rethink television in the context of financialization, commercialization, and vernacular globalisms, such that NDTV's *The Big Fight* and the like will be recognized as a locally produced framing of the same, an appearance of television doing ideological work at home and abroad, for home and abroad. Second, recognize the residues of nationalist and national construction project television, and cinema, in the context of geopolitical reorientation as well as a neoliberal vernacular that trades on a globe-facing specificity, where local incidents articulate with international themes, shaped by commercialization, fianancialization, and neoliberal political alignments. Third, develop a theory of attention and attraction in the media comportment of the news channel that shapes both the globe-facing home audience as well as the occasional incident-related global scrutiny that comes from outside—and do this with considered attention to the labour that goes into framing news as entertainment and as ideology. If I were a little more confident of grand projects, I would suggest that it is worth attempting to develop these three points as a new theory of the media without media theory, as a critique of fascism without fascists, of combating terrors without terror, and as a way of learning to watch television in order to see it for news.

Notes

1. This is a remix version of a developing critique of 'terrorism' on television that is meant as a complement to my earlier books, *Rumour of Calcutta* (on tourist versions of India) and *Critique of Exotica* (on the commercializing of South Asian music), insofar as a rumour and a lurid exoticism organizes media reception of and reflection upon ideological fallout from recent geopolitical shifts. This was always something in which we, I, you, us, all had a great portion of complicity. However much it suits certain interests to present Muslims/Asians/Others (the different categories merge inexorably) as terrorist, and however much 'debates' might challenge or reinforce these moves, the provocations and reactions, the contradictions and ambivalences, the calamities and the contexts, swirl around in a frame that requires us to move elsewhere—beyond media and terror—if we are to make sense of its systemic characteristics. The text was first presented in a quite different form as the keynote of the Sacred Media Cow conference at School of Oriental and African Studies (SOAS)

in October 2007, and then appeared in the proceedings of that conference as John Hutnyk, 2011, 'NDTV 24×7: The Hanging Channel', in Somnath Batabyal, Angad Chowdhry, Meenu Gaur, and Matti Pohjonen (eds), *Indian Mass Media and the Politics of Change*, London: Routledge. It was subsequently refined and remixed for the Shimla conference, '50 Years of Indian Television: Contemporary Issues', held in July 2009, and benefited immensely from the discussion there. Thanks to all concerned. Errors that remain are, of course, mine.

2. Available at http://www.justiceforafzalguru.org/actions/ai.html.

3. The repetitions of this farce multiply with regular recourse to stories about Afzal himself wanting to be martyred, wanting to die, wishing it was over. Another example in May 2010 quotes Afzal saying that his hanging would not cause communal problems (Datta 2010). Curiously, we also have news that he is reading books by communist ideologue Manabendra Nath Roy: 'He is reading Politics, Power & Parties and New Humanism both by MN Roy', according to his lawyer, N.D. Pancholi. An archive of materials from the case is available at http://justiceforafzalguru. org/ (accessed on 24 August 2010).

4. Available at http://www.ndtv.com/video/player/news/merciless-long-queue-for-mercy-petitions/36921 (last accessed 12 October 2012). The author laments, as everyone reading this must know, that a justice system of such machinations does not bode well for the prospects of someone like Afzal, whose life was snuffed out by state intervention in the anticipated, bureaucratic, juridical, but not just, way. On 3 February 2013 President Pranab Mukherjee rejected all mercy pleas and six days later the victim was hanged and buried in the jail grounds with no prior warning to his family. Several days of protests in Kashmir followed.

5. Available at http://www.ndtv.com/video/player/walk-the-talk/walk-the-talk-with-j-k-dutt/66590 (last accessed on 12 October 2012).

6. And in 2010, as an 'app' on IPAD.

7. For Badiou, see the *Hard Talk* interview broadcast on 24 March 2009.

8. Broadcast on 18 September 2008.

9. Available at http://www.countercurrents.org/kavita250407.htm (accessed on 20 July 2010).

References

Beller, Jonathan. 2006. *The Cinematic Mode of Production: Attention Economy and the Society of the Spectacle*. New Haven: University Press of New England.

Bharucha, Rustom. 1998. *In the Name of the Secular: Contemporary Cultural Activism in India*. New Delhi: Oxford University Press.

Derrida, Jacques and Bernard Stiegler. 2002. *Echographies of Television: Filmed Interviews*. Cambridge: Polity Press.

Dutta, Anshuman G. (2010). 'My Hanging won't Create Any Problem: Afzal Guru', *Mid Day*, 27 May, available at http://www.mid-day.com/news/2010/may/270510-Afzal-Guru-Delhi-government-mercy-plea-Hanging.htm (accessed on 1 August 2010).

Geelani, Syed Bismillah. 2006. *Manufacturing Terrorism: Kashmiri Encounters with Media and the Law*. New Delhi: Promila & Co.

Gupta, Nilanjana. 1998. *Switching Channels: Ideologies of Television in India*. New Delhi: Oxford University Press.

Kaur, Raminder and Ajay J. Sinha. 2005. *Bollyworld: Popular Indian Cinema through a Transnational Lens*. New Delhi: Sage Publications.

Kureishi, Hanif. 2008. *Something to Tell You*. London: Faber and Faber.

Mankekar, Purnima. 1999. *Screening Culture, Viewing Politics: An Ethnography of Television, Womanhood, and Nation in Postcolonial India*. Durham: Duke University Press.

Mehta, Nalin. 2008. *Indian Television: How Satellite News Channels have Changed the Way We Think and Act*. New Delhi: HarperCollins.

Morley, David. 1980. *The Nationwide Audience: Structure and Decoding*. London: BFI.

Nandy, Ashis. 1989/2000. The *Tao of Cricket: On Games of Destiny and the Destiny of Games*. New Delhi: Oxford University Press

Prasad, Madhava. 1998. *Ideology of the Hindi Film: A Historical Construction*. New Delhi: Oxford University Press.

Rajadhyaksha, Ashish. 2009. *India in the Time of Celluloid: From Bollywood to the Emergency*. Delhi: Tulika Books.

Rajagopal, Arvind. 2001. *Politics after Television: Hindu Nationalism and the Reshaping of the Public in India*. Cambridge: Cambridge University Press.

Ronell, Avital. 2010. *Fighting Theory*. Urbana: University of Illinois Press.

Roy, Arundhati (ed.). 2006a. *The December 13 Reader: The Strange Case of the Attack on the Indian Parliament*. Delhi: Penguin Books.

———. 2006b. *The Guardian*, 15 December. Available at http://www.the-guardian.com/world/2006/dec/15/india.kashmir (last accessed 12 October 2012).

Spivak, Gayatri Chakravorty. 1999. *Critique of Postcolonial Reason: Towards a History of the Vanishing Present*. Cambridge, MA: Harvard University Press.

Spivak, Gayatri Chakravorty. 2000. 'Discussion: An Afterword on the New Subaltern', in P. Chatterjee and P. Jeganathan (eds), *Subaltern Studies X1: Community, Gender and Violence*, pp. 305–34. New York: Columbia University Press.

———. 2008. *Other Asias*. Oxford: Blackwell Publishing.

Stiegler, Bernard. 2009[1996]. *Technics and Time, 2: Disorientation*. Stanford: Stanford University Press.

Thussu, Daya Kishan. 2007. *News as Entertainment: The Rise of Global Infotainment*. London: Sage Publications.

Vidal, Gore. 2000. *The Golden Age*. London: Little Brown.

Virdi, Jyotika. 2003. *The Cinematic ImagiNation: Indian Popular Films as Social History*. Delhi: Permanent Black.

Wayne, Mike. 2003. *Marxism and Media Studies*. London: Pluto Press.

9

Big Brother, Bigg Boss

Reality Television as Global Form

BISWARUP SEN

Reality programming is 'bigg' with the Indian television audience. Consider *Bigg Boss*, the Indian version of *Big Brother* that has aired for seven seasons on the Colors channel. The show has grown in ratings every successive outing, and with the finale of its fourth season, it pulled in the astonishing figure of 6.7 Television Viewer Ratings (TVRs).[1] To put this figure in perspective: in the last week of June 2011, the top-rated drama show on Colors, *Balika Vadhu*, had a TVR of 3.81, and the most popular show across all channels, *Pavitra Rishta* (Zee TV), had a TVR of only 4.02.[2] This trend is here to stay. The appearance of Bollywood superstars, Sanjay Dutt and Salman Khan, as co-hosts for season five, the 8.1 TVR that *Bigg Boss 4* earned with higher-income groups, and the recent finding that 76 per cent of Indian youth between the ages of 12–18 years prefer reality programming over any other kind, collectively suggest that *Bigg Boss* and its variants will be a dominant feature of the Indian television landscape in the years ahead (Indiantelevision. com. 2011; *The Telegraph* 2011). This chapter contends that reality TV's enormous popularity in India—and elsewhere—represents what may be termed as the *globalization of the aesthetic*. That a format like *Who Wants to be a Millionaire* plays in 106 countries is testament to the fact that reality TV is the most universal form of television programming that exists today. Its widespread appeal cannot be explained away as passing fad or marketing strategy, but must be seen as an index of the *real* processes that constitute globalization. Consequently, an inquiry into the appeal and meaning of reality TV is likely to yield important insights into the nature of contemporary global culture.

In this chapter, I use one well-known reality format—*Big Brother* (*Bigg Boss* in India)—in order to shed light not only on a unique television phenomenon, but also on the complicated relationship between the forces of globalization, national and local cultural formations, and the dictates of commercially driven entertainment. In the first section, entitled 'The Real as Global', I analyse the essential features of format television to argue that the very mode of its constitution as an economic and aesthetic object inclines it towards the global. In other words, the logic of reality TV is formally equivalent to the logic of globalization. In 'The Global as Real', I try to elucidate the reality behind global cultural formations by discussing the two main theoretical approaches to the question of global culture: cultural imperialism and cultural globalization. Whereas cultural imperialism saw globalization as a linear process of takeover and homogenization that reduced all national cultures to the level of McDonald's and Coca Cola, cultural globalization argues for the potency and even the primacy of the local and the particular in determining the outcomes of global cultural flows. I offer a reading of reality TV in the light of these paradigms and suggest that neither of them fully exhaust the possibilities of the form. In my concluding section, I offer some speculations about how reality TV embodies global form and thus functions as a sort of 'Bigg Boss' that dictates contemporary modes of meaningful behaviour.

The Real as Global

It is not entirely coincidental that reality programming—like many other forms of contemporary programming—was made possible by a global event. In the words of a prominent scholar of the medium, 'what used to be one of the most closed broadcasting systems in any democracy' was forced to open up when Indian audiences began to tune in to CNN's live coverage of the Gulf crisis (Thussu 1998: 277). Within the span of a few years, television in India was multichannel, multinational, and multifaceted. Whatever one's views on this passage from 'monopoly to polyphony',[3] it had two undeniable consequences. First, it brought all programming under the imperative of profit maximization. Henceforth, television would no longer be considered an instrument of progressive social change; its main purpose now would be to entertain and generate advertising revenue—a change in broadcasting philosophy dramatically acknowledged in the mid-1990s by Doordarshan's decision to include *Dallas* and *Dynasty* on its Metro channel lineup (Thomas 2005: 103).

The opening of the skies also made possible the direct importation of global television content. Channels such as CNN, CNBC, Discovery, ESPN, and MTV, as well as regional networks like STAR TV and Zee TV, made available a wide range of international programming—Hollywood movies, sports coverage, news, soaps and serials, informational shows—that the domestic viewer had been unaccustomed to. More importantly though, the new ethos encouraged Indian television producers to imitate and improvise on global styles of television making. The first notable 'software transfer'[4] happened as early as 1982, when the then Information and Broadcasting Minister, Vasant Sathe, was inspired by Mexican telenovas to come up with the idea for *Hum Log* (1984–5), the first truly successful serial to air on Indian television (Chatterji 2008). A couple of decades later, the Colombian soap opera, *Yo soy Betty, la fea*, would serve as the model for the equally successful, *Jassi Jaissi Koi Nahi* (2003–6). Yet, the 'real' global invasion, I want to argue, happened with the advent of reality TV as a worldwide phenomenon. Consider the following:

1. *Bigg Boss*, 2006 (original version: *Big Brother*, the Netherlands, 1999; American version: *Big Brother*, 2000; owned by: Endemol).
2. *Indian Idol*, 2004 (original version: *Pop Idol*, the United Kingdom (UK), 2001; American version: *American Idol*, 2002; owned by: FreeMantle).

3. *Kaun Banega Crorepati*, 2000 (original version: *Who Wants to be a Millionaire*, the UK, 1998; American version: *Who Wants to be a Millionaire*, 2002; owned by: Sony Pictures Entertainment).
4. *Khatron Ke Khiladi*, 2008 (original version: *Now or Neverland*, the Netherlands; American version: *Fear Factor*, 2001; owned by: Endemol).
5. *Jhalak Dikhla Jaa*, 2006 (original version: *Strictly Come Dancing*, the UK, 2004; American version: *Dancing With the Stars*, 2005)
6. *MasterChef India*, 2010 (original version: *MasterChef*, the UK, 1990; American version: *MasterChef USA*, 2000, *MasterChef*, 2010; owned by: Union Pictures Production).
7. *Sacch Ka Saamna*, 2009 (original version: *Nada más que la verdad*, Colombia, 2007; American version: *The Moment of Truth*, 2008; owned by: Lighthearted Entertainment).

There are three features of this list are that worth noting, each of which alerts us to the strong relationship between globalization as process and reality TV as cultural form. First, consider the extreme rapidity with which reality formats are implemented across the globe. *Pop Idol* was 'invented' in 2001; it took a mere three years for its Indian equivalent to begin airing. This rapidity signals a break with a temporal regime in which cultural products travelled slowly (if at all) from centre to periphery. The notorious 'lateness' of postcolonial time is replaced by what David Harvey described as 'space–time compression', by virtue of which everything happens everywhere simultaneously. In this new topology, culture can loop in reverse—thus, *Millionaire* airs in India before it does in the United States (US), *The Moment of Truth* debuts in Colombia before any place else.[5]

Second, reality TV proliferates at astonishing speeds. For example, the number of reality shows on Indian television has grown at an exponential rate in less than a decade: apart from the titles listed earlier, one could mention *Roadies*, *On the Job*, *The Great Indian Laughter Challenge*, *Rakhi Ka Swayamwar*, *Splitsvilla*, *Jodi Number One*, *Voice of India*, *Emotional Atyachar*, *Dadagiri*, *Big Switch*, *The Chair*, and many more. This epidemic aspect of the form is brought out even more strikingly in the American context—*Reality TV Magazine* provides daily updates on over 200 reality shows, while the Wikipedia entry entitled 'List of reality television shows' is so long as to seem non-denumerable.[6] These figures are so large when compared to the number of crime dramas or sitcoms that have originated in the same time period that one must conclude that we are in the presence of a very different sort of cultural production here. This astonishing rate of reproduction is enabled by reality TV's intrinsic ability to *clone itself with a difference*. Today, *Extreme Makeover*, tomorrow *Extreme Makeover Home Edition*. This attribute is not specific to reality TV, but rather an abstract principle or force that is central to globalization as well. If globalization is understood as a process that is based on sameness-and-difference (apply the same free market principles to obtain different outcomes), we begin to perceive the strong resemblance between its *form* and that of reality TV.

Finally, we must take heed of the entries listed under the rubric of 'Original Version'. Four of the shows in the list—*Pop Idol, Who Wants to be a Millionaire, Strictly Come Dancing, MasterChef*—originated in the UK; two—*Big Brother* and *Now or Neverland*—in the Netherlands; and one—*The Moment of Truth*—in Colombia. It is impossible not to notice the glaring absence of the US from this list. During the course of the 'The American Century', most media content that was globally significant originated in the US: think of Hollywood, *I Love Lucy, Dallas*, jazz, rock and roll.[7] What, then, are we to make of the lack of American enterprise in the case of reality TV? It is possible to approach this anomaly in purely economic terms. Scholars like Chad Raphael and Ted Magder have argued that reality TV arose as a response to the growing costs of traditional programming as well as the fracturing of the television audience due to the introduction of cable. Yet, as Magder (2009) himself argues, there is more to reality TV than just the attempt to cut production costs and financial risk. He goes on to identify four significant changes to the production side of television that it illustrates: the use of prepackaged formats; product placement; multimedia exploitation of television programmes; and the growing strength of European programme suppliers to the American and international television market. Magder (2009: 150) explains the last development by pointing out that the use of 'unfinished programming' makes 'good sense for a company like Endemol, based as it is on a small national and linguistic market'. Magder's argument makes sense but needs to be complemented by the observation that the replacement of *Dallas* by *Big Brother* as prototypical global entertainment is a cultural as well as an economic fact.

The shift to other centres of the global North may be read as a trivial realignment of the same asymmetric power structure (whereby content and formats continue to flow from North to South), but such a reading would be overlooking two important considerations. First, both the UK and the Netherlands are, as a result of their history, geography, and political and economic ties, far more global in their make-up than twentieth century America. Cultural products originating in more global spaces are, tautologically almost, likely to be more global in nature. Compare English football as it was in 1950 with its contemporary version. In the 1950s, Britain was an island nation and its football was highly parochial in nature—played by Englishmen for Englishmen. Today, Britain is a global financial centre and not surprisingly, the English Premier League is a completely transnational product that is marketed across the globe. This is even truer in the case of reality TV. Thus, *Pop Idol*—the original version of the *Idol* format—was created by an Englishman (Simon Fuller) and first aired on a British television channel (ITV). However, the fact that the *Idol* format plays in 42 countries and that it is owned by FreeMantle, a global entertainment company largely owned by the giant German conglomerate Bertelsmann, makes any assertion about its intrinsic Englishness quite meaningless.

To recap, the speed of dissemination, the profusion of different formats, and the weakness of origin, all point to a very specific mode of production (and distribution) that creates a mobile cultural form that can proliferate worldwide. However, we need to look beyond the economic determinants of reality TV and look also at its internal structure—its 'mode of constitution' as it were—in order to fully appreciate its global dimensions. If there is one feature that distinguishes reality TV from all other previous forms of television programming, it lies in the fact that every reality show is based on a format. That is, below the level of content consumed by the viewer, there is a more primitive level that can be thought of as a set of rules and instructions that allow a reality show to 'happen'. This suggests that formats are novel objects that cannot be entirely subsumed under the notion of the 'artwork'. A crucial difference between the two lies in their relation to semantics. The traditional work of art is characterized by the fact that it is suffused, indeed overloaded, with meaning— calling for elaborate hermeneutics and endless interpretation. The semantic content of the artwork has traditionally been thought of as independent of our interaction with it. There are, of course, competing accounts of what this pre-existing meaning consists of: the author's intention; his or her biography; the socio-historical background or context; the formal features of the text; or its psychological and psychoanalytical substratum. In recent years, some theorists have tried to introduce what could be clumsily termed as 'after-semantics', claiming that the literary work as a meaningful entity arises out of the interaction between text and audience. For reception theory, the literary work emerges as a third term situated between the text and the reader. A more sociological version of this doctrine—which replaced the individual reader with groups of readers—was developed within the cultural studies tradition in the work of Stuart Hall (encoding/decoding), David Morley (audience studies), and John Fiske (resistance).[8] While both these approaches render the text polysemic, it is doubtful whether they succeed in relocating meaning from its traditional address. However variously different communities of viewers interpret *Dallas*, it strains one's credulity to argue that the work-in itself is semantically empty (meaningless) and acquires meaning only after it is viewed. In short, the artwork and its meaning are sutured together and it is impossibly hard to separate one from the other.

Formats, on the other hand, are woven by a syntactical thread and carry the lightest of semantic loads. The 'playbook' for *Big Brother* reads roughly as follows: *choose 12 contestants; put them in a house along with a host; assign some tasks; arrange some contests; institute a policy that results in one contestant being eliminated every week; arrive at the eventual winner by this process of elimination; repeat series with a new set of contestants.*

This sequence is not so much a text as a set of rules, instructions, or codes that is inserted into different initial conditions in order to produce a text—*Big Brother Australia* or *Bigg Boss*. Formats, then, are 'thin' texts whose semantic value is close to zero. Unlike the artwork, they do not have meaning; instead, they have the potential to have meaning, to become texts. This fundamental property—a 'bare semantics' if you will—explains why, in the words of one industry professional, formats 'travel without being stopped by either geographical or linguistic boundaries' (Moran 2006: 27).

I want to examine two specific features of formats—authorship and narrator/host—in order to demonstrate how the shift from semantics to syntax enables reality TV to attain a high 'global index'. The semantic poverty of reality formats makes the question of authorship and origin quite moot. The dissociation between author and work in the case of reality TV is far more radical, thoroughgoing, and definitive than that envisioned by the 'death of the author' thesis postulated by thinkers like Barthes (1977) and Foucault (1984). Whereas poststructural theory sought to dissolve the author (and authorship) in a discursive 'solution', this theoretical effort failed in practice: contemporary classics like *Pulp Fiction* or *Beloved* cannot be contemplated without invoking Quentin Tarrantino or Toni Morrison. Yet, very few, if any, of *Indian Idol* or *Bigg Boss*' viewers are either aware or would care to know about Simon Fuller or John de Mol—the original 'authors' of the formats. In other words, because authorship occurs at the syntactic level, its imprint on the work is almost invisible. Devoid of parochial origins, such 'author-free' works are the perfect vehicle for travel across cultural boundaries.[9] The narrator/protagonist central to both the novel and cinema is replaced in reality shows by a new literary type who is best described as host/facilitator. For some reality formats—typically talent shows—the facilitator figure is complemented by one or more judges. Host/facilitators and judges must be seen as syntactical rather than semantic figures, in the sense that their interaction with contestants is an enabling of a procedure rather than the 'meaningful' exchange commonly found between characters in a novel or film. Hosts engage with contestants in a purely operational manner: they inform, assign, allocate, choose, or eliminate; and never emote, get involved, or determine the course of action. The extremely sparse role of the host means that the slot can be filled in variety of ways dependent on the locale that the reality format in question is implemented in. When *Big Brother* plays in America, it features a relatively unknown actress—Joan Chen—as host.[10] Such a strategy is not possible in India, where the contestants are not ordinary folk but are minor celebrities from the entertainment world. Why this is so is not easy to explain; I will only speculate that having such a cast allows *Bigg Boss* (and similar shows) to perform a radical redefinition of subjectivity under the cover of celebrity behaviour. In any case, this situation—and the mandate of compulsory intertexuality with Bollywood imposed on all areas of Indian popular culture—necessitates that the show's host have high recognition value—hence, Shilpa Shetty as host in Season 2 and superstars Amitabh Bachchan and Salman Khan in Seasons 3 and 4 respectively.

The main point to note here is the following. Since the 'host' is a syntactic placeholder, it is a variable that acquires a value depending on the context the format is run. This variability allows reality TV to fill its space with 'global matter', while maintaining the format's logic. One could tentatively extend this observation from the host/facilitator to the contestants themselves. It could be argued that the 'characters' who populate the show are not real three-dimensional humans but rather themselves stock character types—the aggressive one, the betrayer, the conciliator, etc.—who also function as blank slots to be filled in by local content. In other words, I am suggesting that even those who ostensibly look human (and semantic) are, in fact, syntactic counters who will be moved around in the game to produce certain patterns. This post-humanist reading is strongly suggested by the standard opening shots that depict players arriving at the game venue (house, island) from different geographical locations. This is in stark contrast to standard literary or cinematic practice—where many characters in a typical novel or film are posited as 'knowing' each other from an unrepresented past (that is, anterior to page one or the opening credits). The fact that all contestants are complete strangers to each other implies that their mutual interaction is far more determined by the rules of the game rather than any sort of semantic relation. The elaborate emotive strategies employed by producers—the construction of teams or tribes; the making and unmaking of alliances; the encouragement of bad behaviour; hints of romantic interests— are merely superficial and largely a concession to the traditional audience. Their main import is to occlude the truth that at bottom, players are syntactical and not semantic entities on whom the show employs a combinatory logic in order to produce a set of outcomes (elimination, immunity, winner). In this respect, reality shows resemble sports and not the narrative arts—ending with a set of objective results rather than dramatic closure.

To sum up then, reality TV's mode of production—the generation of large numbers of formats by transnational companies that are distributed internationally—as well as its mode of constitution— the primacy of syntax over semantics—work together to create a mode of entertainment which travels effortlessly across the globe. Indeed, reality TV can be considered a minor version of globalization itself. Formats work as a heuristic because they stand in a fractal relationship to globalization: the former involves the running of a few universal rules on local game conditions; the latter consists of the administration of a few macroeconomic prescriptions (free market philosophy, International Monetary Fund [IMF] regulations, etc.) on an entire country. Reality formats give birth to texts when inserted into individual cultures; globalization, in a similar manner, produces a text—not at the level of representation but at that of history. This equivalence has an epistemological pay-off: the fate of formats can serve as a useful model of the fate of globalization itself. The formula here is *Big Brother Africa* = Africa + globalization. Reality formats can therefore provide us with a finely calibrated account—a calculus—of the impact globalization makes in different contexts.

I have established that reality TV is the most ubiquitous form of global programming because it is structurally and operationally designed as a transferable 'technology'. Further, as I have just argued, it mimics the logic of globalization on a miniature scale. In the section that follows, I treat these conclusions as given and go on to discuss the ways in which we can understand the import and significance of reality TV's global spread.

The Global as Real

What is the truth behind the circulation of media products on a globalized scale as instanced by the case of reality TV? For some, this flow recapitulates history and is tantamount to a new sort of imperialism. As an early proponent of this thesis put it, media or cultural imperialism was 'The process whereby the ownership, structure, distribution or content of the media in any one country are singly or together subject to substantial external pressures from the media interests of any other country or countries without proportionate reciprocation of influence by the country so affected' (Boyd-Barrett 1998: 165–6). Such a view received its most succinct expression in the work of Herb Schiller (1971), and as the title of his influential *Mass Communication and American Empire* makes abundantly clear, there was little doubt as to which country was the source of 'external pressure'. The cultural imperialism thesis made popular by Schiller (and others) can be briefly summarized as follows. The media industries are a crucial part of modern capitalist societies: they produce audiences for advertising; they generate huge revenues; and finally, they function as a 'consciousness industry' that legitimizes and sells the capitalist way of life. Since capitalism is inherently expansionist, it is only to be expected that the advanced capitalist countries of the West would embark on a systematic mission to 'colonize' media systems in the developing world in order to make these cultures import Western programming and thus convert to the 'American way of life'.

Viewed from this perspective, the global spread of reality TV fits right in with the logic of capitalist production and expansion. Reality shows can be cheaply produced in great numbers, are easy to export and implement in other cultures, and are thus able to earn revenue in markets that had hitherto been hard to penetrate. There is no contemporary work, to my knowledge at least, that specifically looks at reality TV from a cultural imperialism framework. This lack is understandable given that the perspective has largely gone out of fashion, and is more often than not a benchmark *against* which people define their own views. It is possible, however, to extrapolate from the writings of a political economist like Robert McChesney and come up with version of what such a reading might look like. In his well-known piece, 'Global Media, Neoliberalism and Imperialism', McChesney (2001) points out that whereas previously media systems were primarily national, the last few years have seen the emergence of a global commercial media market. Thus, a few multinational corporations like Disney and Bertelsmann, as well as a few dozen second-tier national or regional powerhouses, dominate the global media marketplace. This set-up has been made possible by the dominant ideology of neoliberalism, which refers to the set of national and international policies that call for business domination of all social affairs with minimal countervailing force. The media environment that is created by this regime is so thoroughly global that it can no longer be considered a purveyor of the US culture; rather, the 'global media system is better understood as one that advances corporate and commercial interests and values and denigrates and or ignores that which cannot be incorporated into its mission' (McChesney 2001: 16). Reality TV can then be read as the perfect exemplar of such a global media system insofar as its mode of production—as I have discussed earlier—as well the nature of its reception are highly internationalized. The remaining challenge for such a reading would be to demonstrate that reality TV is structured by neoliberal thinking and that it 'advances corporate and commercial interests'. Such analyses do exist—in recent work by Andrejevic (2004) or Ouellette and Hay (2008), for instance—but they cannot properly be located within the political economy of media perspective that I have been discussing.[11]

The cultural imperialism approach has, for some time, been the target of strong criticism. In a well-known piece, John Tomlinson (1997) argued that it is not possible to make sense of globalization in the manner suggested by the cultural imperialism thesis because of three interrelated reasons. First, the evidence of the distribution of and impact of Western/American goods is more ambiguous than one may expect. To take two examples discussed by other authors: after an initial phase of heavy American imports, the Latin American television market was dominated by regional products from Mexico and Brazil (Cunningham et al. 1998: 179); again, in the Asian case, we find that globalization has led to the pluralization of cultural production as evidenced by the dramatic growth in the Hong Kong film industry (Bannerjee 2002). Second, we need to conceptualize cultural processes in dialectical terms and realize that cultural influence is unlikely to follow the sort of linear paths that cultural imperialism predicts. Finally, *pace* cultural imperialism, the globalization process is a decentred one that does not map neatly on to the geographies of domination established in the era of colonialism.

What conclusions follow when we study reality TV from a perspective such as Tomlinson's? As the editor of a recent volume on global perspectives on reality TV puts it:

Rather than positioning the global spread of reality television formats as another example of US or even Anglo cultural imperialism...new work suggests that the relationships among global formats and their local implementations are specific, contingent and unpredictable. Each nation that develops and adapts reality formats does so in a particular policy environment that navigates between (at least) commercial demands, expectations about public interest programming and local tastes...The direction of flow of formats, as well as their adaptability, challenges the assumption of US cultural imperialism in reality TV... (Sender 2011: 4)

In other words, though reality formats are mainly developed in the West (with some exceptions like *Iron Chef*, which was developed in Japan), they are not 'Western' insofar as their implementation necessarily involves the intervention of local conventions and practices. What you get after you implement a format in a specific country is a compromised product unique to that media environment. Consider the wide range of outcomes in the case of *Big Brother*:

1. The first British series of *Big Brother* included contestants stripping off their clothes, and in Holland, the contestants were even more uninhibited than in Britain. By contrast, the American *Big Brother* contestants talked a lot about sex and relationships, but remained modestly clothed and no sexual liaisons took place (Bignell 2006: 49).
2. Instead of presenting a summary of daily events in the house as is done elsewhere, the Brazilian producers of *Big Brother*

started to develop a hybrid language that mixed reality TV with soap opera, a move that created record-breaking profits for Brazil's largest broadcaster, Globo (Campanella 2009).

3. *Big Brother* Australia performed its 'Australian-ness' in a variety of ways. According to the network executive in charge of the show, the house that was used for the show was very different from those in overseas versions because 'we wanted this to be a real Aussie house, that means relaxed lifestyle, sunshine, backyard pool, backyard BBQ, a real Aussie *Big Brother'*. Again, there was a far greater emphasis on fitness and outdoor activities in the Australian version. Finally, unlike the American *Big Brother*, contestants remained indifferent to the outcome of challenges. Whereas in many countries, challenges have played an important role in generating some of the emotional drama and conflict, in Australia, 'it's a cultural thing...They don't seem to give a bugger whether they win or lose' (Roscoe 2004: 312–13).

4. The original Dutch version had situated participants in a 'wealthy' house and send losers to a stable as the 'poor house'. This was not considered funny but rather as offensive in the Argentine context, given the country's dismal economic condition at the turn of the millennium. Moreover, local producers frowned upon the use of house that featured a swimming pool, gourmet kitchen, etc., on the grounds that it was typical of wealthy rather than average people (Waisbord and Jaslin 2009: 67).

5. *Al-Ra'is*, the Arabic version of *Big Brother*, was shut down after a week following a huge controversy that was started by an on-screen kiss. For Islamists, the show (which, interestingly, only featured women who were divorced) violated Muslim values by putting unmarried men and women in a confined physical space. Another Endemol format, *Star Academy*, was described as 'Satan Academy' and had a fatwa issued against it (Kraidy 2009: 33).

These instances suggest the slogan: *no globalization without localization*. The 'cultural globalization' thesis—which is by any measure the dominant approach in the field of global media studies today—is based on this intuition and makes two strong claims about media, culture, and society in the age of globalization:

First the flow of global culture is not one-directional as assumed by the cultural imperialism; rather, as famously described by the anthropologist Arjun Appadurai (1996: 32, 42): 'The new global cultural economy has to be seen as a complex, overlapping, disjunctive order that cannot be understood in terms of existing center–periphery models...The globalization of culture is not the same as homogenization, but globalization involves the use of a variety of instruments of homogenization...that are absorbed into local political and cultural economies.' In other words, cultural exchange today is a multilateral process that produces diversity and difference.

Second the output of this interaction or articulation is a complex entity that is captured by such terms as 'hybridity' or 'glocalization' (see Kraidy 2005; Robertson and White 2007). The hybridity model—first enunciated by theorists like Homi Bhabha and Nestor Garcia-Canclini—sees international culture as emerging from a struggle for dominance between global and local formations. Glocalization theory, on the other hand, operates with a less antagonistic view of this encounter contending that 'Rather than speaking of the inevitable tension between the local and the global it might be possible to think of the two as being different sides of the same coin' (Robertson and White 2007: 62).

The appeal of the cultural globalization approach is enhanced by the fact that its theoretical suppositions are in tune with the major intellectual trends of the past few decades. It is clear that the hybridity thesis derives from cultural studies—the model of articulation between dominant and oppositional forces that it employs is exactly like that used earlier to analyse cultural politics in the domestic context. Cultural globalization is also broadly in tune with postmodernist thinking, and thus it scores over its rival because 'in the process by which postmodernism has succeeded neo-Marxism as the master paradigm in social and cultural theory, the new orthodoxy has taught us to be skeptical of such "grand narratives" or totalising theories as that of cultural imperialism' (Cunningham *et al.* 1998: 180). Finally, cultural globalization also draws upon the notion of the 'active audience', proposed by reception theories that suggest that 'viewers are capable of making interpretations in many different ways...and thus the meaning of TV imports is always subject to local contexts of reception' (Bielby and Harrington 2008: 25).

Even a cursory look at *Bigg Boss* (the Indian version of *Big Brother*) would give strong support to the cultural globalization hypothesis. Though *Bigg Boss* follows the standard rules of the format—14 contestants reside in a house located in Lonavala, a tourist resort near Pune, Maharashtra, and are ejected one by one based on peer nominations as well as audience voting in order to determine the final winner—one can enumerate a number of ways in which it is hybridized or glocalized:

1. The introduction of Bollywood figures as hosts and guests. Thus, *Bigg Boss 3* was hosted by the superstar Amitabh Bachchan and *Bigg Boss 4*, by Salman Khan. *Bigg Boss 4* also featured a long list of Bollywood notables—Ajay Devgan, Kareena Kapoor, Farah Khan, Rani Mukherjee, Katrina Kaif, Vidya Balan, Dharmendra, and Bobby Deol—who appeared on the show to promote their current work. This 'invasion' reflects a peculiar Indian tendency whereby every form of media and entertainment is inflected by Bollywood's influence.
2. India's obsession with its political Other is reflected in the inclusion of *two* Pakistani artists—the actress, Veena Malik, and the cross-dressing performer, Ali Saleem—in the cast of 14.
3. While some the tasks assigned to the inmates are culturally unspecific—thus Dixcy Scot (the week three task in *Bigg Boss 4*) involved wrestling matches between male inmates, while females functioned as cheerleaders—some others are quite unmistakably Indian in form and content. Relevant examples from *Bigg Boss 4* are: *Shakti De Bhakti De Mukti De* (week two: contestants wear a white gown and chant to a particular tune); *Gaon Gaon Shahar Shahar* (week seven: some inmates have to live like villagers, while others have to live like city folk); and *Jee Lay Jee Jaan Se* (week 13: each

housemate is given a specific song everyday and they have to dance to it.

4. As mentioned earlier, in keeping with India's newfound celebrity culture, all the contestants are well-known media figures. Thus, the fourth Season of *Bigg Boss* featured Abbas Kazmi, a famous criminal lawyer; Seema Parihar, a former woman bandit; Dalip Singh Rana, a professional wrestler; and the model, Sara Khan. This is in contrast with, say, the American version, where housemates are unknown and presented as 'average' people.

These examples push us towards the conclusion that as in every other part of the world, reality TV in India illustrates what is perhaps the crucial formula behind the production of global culture: the insertion of local content into international formats. The consequence of this strategy, according to some, is a healthy increase in cultural diversity. Thus, two eminent scholars of global formats argue, 'program ideas are now more diverse across nations...a deepening of international trade in television programs has contributed positively to diversity. Indeed, in many Asian countries the TV landscape is more multicultural and cross-cultural than a decade ago' (Keane and Moran 2008: 155). In other words, however universalizing or homogenizing the tendencies of capital, the ultimate primacy of the local ensures a proliferation of difference.

* * *

While acknowledging the contribution and significance of the cultural globalization thesis, I want to, in this concluding section, argue against its wholesale acceptance and also speculate about other ways of conceptualizing the fundamentals of global culture. The three shortcomings of the dominant approach that I want to draw attention to are as follows. First, cultural globalization has a natural bias towards 'methodological nationalism', that is, of treating all issues of culture and global politics within the boundaries of the nation-state. Such a way of thinking may blind us to the *global* possibilities of culture as opposed to the national manifestation of global forces. As Kevin Robbins and Asu Aksoy (2005: 21) have eloquently argued:

> The key issue in this respect, it seems to us, is whether we can find new ways that work against the gravity field of the national imaginary. A different kind of media order—a transnational or transcultural order—cannot come into existence automatically. We will have to think it into existence, thinking counter-nationally, thinking against the grain of the national mentality.

A similar call is issued by the German sociologist, Ulrich Beck, when he calls for a cosmopolitan sociology that will re-envision the national as the *internalized global* (Beck 2002). Most work in the cultural globalization tradition acknowledges the global but studies it only in the context of a specific local culture. Thus, we get numerous studies that look at reality TV in places like Malaysia or Slovakia, do an excellent job of explaining the nuances of local cultures and how they affect the execution and reception of reality formats but give little thought to the global dimensions of the form (see, for example, Barrer 2010 and Wahab 2010). How do we redress this imbalance? Robbins and Askoy (2005) propose thinking 'counter-nationally' by means of the categories of 'experience' and the 'experiencing subject'—their own work looks at how Turkish migrants based in London use Turkish-speaking satellite channels to produce a diasporic subjectivity that is resolutely transnational. Such a project need not however be restricted to phenomenological inquiry; an analysis of aesthetic objects (as I am proposing next) would also provide us with clues as to the nature of global form.

Second, the implicit Manichaeism in the binary global/local leads to an impoverished conception of the global. The global is typically seen as invasive, intrusive, and imposing a set of alien values and practices which are without virtue. The local, on the other hand, is celebrated in the name of resistance, agency, and multiculturalism. Not surprisingly, in this scenario, David *always* prevails over Goliath. The etiology of this perspective is easy to discern: in trying to correct the monolithic narrative of the cultural imperialism thesis, cultural globalization necessarily leans too far in the other direction. The price to pay for this redress is

Channeling Cultures

quite steep, for the more one celebrates the local, the less attention one pays to the category of the global. As a consequence, the notion of the global operational in cultural globalization theory is not, in the last instance, all that different from that used by the cultural imperialism thesis. In both versions, the global is always in harness with capital, is imbued with a culture that is distinctly Western, and has a natural propensity to reduce difference in the name of entertainment and profit maximization. The only significant regard in which these two theses differ is in their accounts of what happens in the encounter between the global and the local. We need a deeper and richer account of the global than what these versions provide, and again, a close examination of reality TV— and similar global cultural forms—would provide us with the necessary tools to construct a more fleshed out account of global culture.

Third, more specifically, the limited view of the global implicit in cultural globalization does not allow for an adequate analysis of reality TV. While work in this tradition has produced rich accounts of the vicissitudes of reality formats in a variety of cultural contexts, the paradigm is less capable of addressing the question: why reality TV? The statistic that 76 per cent of Indian teens prefer reality programming cannot be explained by pointing to the variations on the *Big Brother* format that we find in *Bigg Boss*. We must turn to the essential structure of *Big Brother–Bigg Boss* in order to grasp reality TV's universal appeal. In other words, I am suggesting that we conceptualize reality TV as a sort of *universal machine* that engineers global effects through the mechanism of format implementation. Scholars like Michael Keane, Albert Moran, and Barry King take a step in this direction when they describe reality television as an *engine* that produces a specific performance commodity as well as renown based on talent and ingenuity (Keane and Moran 2008; King 2009). This intuition can be developed further to address other aspects of reality TV as well. I want to close this chapter with a very brief discussion of an example of a global 'universal'—shamelessness. My analysis is obviously neither substantive nor comprehensive; all it seeks to do is to indicate an alternative methodology for studying reality TV.

Most viewers of reality TV would agree that its contestants display a remarkable degree of shamelessness. Whether it is Snooki in *Jersey Shore* or Dolly Bindra in *Bigg Boss 4*, most reality participants are ready to display every sort of emotion, divulge the innermost of secrets, and indulge in a provocative and unrestrained mode of interaction with others. This phenomenon could be analysed in a normative manner as exemplifying the decay of manners and the triumph of bad taste, but a better strategy would be to try and locate shamelessness within the conceptual framework of realty TV and thus, configure its place in the ontology of the global. Shame, as anthropologists have pointed out, is the affect that both indicates and atones for an individual's deviance from societal norms. Shame can be felt because the individual is always under the judging eye of society as Other; it thus presupposes the existence of a social mass both anterior and superior to the individual.[12] The structuring logic of reality TV precludes the formation of such a social mass. As I have alluded to earlier, the artificial society—14 contestants living in the *Bigg Boss* house, for example—constructed by every reality show starts without history (even the imagined history of novels and cinema) or sociology (hierarchies, conventions, groups). The mass of such a society is purely arithmetical, expressed by the sum total of contestants currently in play. For such a complex, judgement is a matter not of a single self being scrutinized by the whole of his fellowmen, but rather one individual being evaluated by a jury of one. To put it a little differently, shame can only occur when individual behaviour is placed against objective norms. What we get in the reality show is a series of individual judgements that have the subjective status of opinions or beliefs. The surfeit of judgement evident in any episode of a reality show occludes the fact that, despite the famous line from *Survivor*, the tribe can never speak.

Thus, the logic of the reality show makes shame an impossible emotion. One begins to see how this very absence may portend the emergence of a new type of subjectivity, one that is both shameless and global. To turn to anthropology again, it is well known that shame is not a context-free category—what is shameful in one culture (baring the bosom, say) is quite irreproachable in another. There is therefore an *a priori* reason to hypothesize that shame cannot exist in the domain of the global. The shedding of all reticence and modesty on Facebook and other types of social media lends weight to the view that shamelessness may be the condition to which we must all aspire in the new digitized global age. Shamelessness is then a kind of 'sub-routine' embedded in the algorithmic structure of reality TV. When such a programme is 'run' in a particular context, it not only acquires the values of locality, but in the same instance, imposes a certain code (of which shamelessness is one instance) that defines the limits of the possible. Such a code, visible in the theater of reality TV, and still in the process of being written, is the right place to discover the meaning of the global.

Notes

1. TVR stands for television ratings. One single television ratings point represents 1% of viewers in the surveyed area in a given minute.

2. Indintelevision.com, available at http://www.indiantelevision.com/ tvr/indextam.php4?id=2138&startperiod=&endperiod= (accessed on 12 July 2011).

3. See Adrian Abbott Mathique (2009) for a concise account of the transition from state-run broadcasting to privatized television. Also, see Usha Manchanda (1998) and Amos Owen Thomas and Keval J. Kumar (2004).

4. In India, as in many other countries, television content is commonly referred to as 'software'.

5. It could also be argued that *Sa Re Ga Ma Pa*, a singing contest that debuted in 1995, was the world's first *Idol* show.

6. Available at http://en.wikipedia.org/wiki/List_of_reality_television_ programs (accessed on 27 June 2011).

7. Bollywood and Hindi film music was a notable exception to this tendency. For an excellent account of Bollywood's transnational impact, see Sangita Gopal and Sujata Moorti (2008).

8. See, for example, Hall (1980), Morley (1980), Fiske (1989).

9. Music impresario created the *Idol* format in 1998. It debuted as *Pop Idol* on ITV in October 2001. Available at http://en.wikipedia.org/wiki/ Pop_Idol (accessed 5 May 2008). The idea for *Big Brother* came during a brainstorming session at the production house of John de Mol Produkties (an independent part of Endemol) on Thursday, 4 September 1997. The first *Big Brother* broadcast was in the Netherlands in 1999, on the Veronica TV

channel. Available at http://en.wikipedia.org/wiki/Big_Brother_%28TV_series%29 (accessed on 5 May 2008).

10. Hosts of successful reality shows do end up being 'known' but in the sort of half-celebrity mode enjoyed by news readers, talking heads, and sports commentators.

11. Ouellette and Hay (2008) use the Foucauldian notion of governmentality to argue that reality TV is instrumental in producing a self-governing neoliberal subject. Andrejevic (2004) analyses reality programming as a device that creates value out of affective labour.

12. Anthropologists also refer to 'guilt' societies where individuals are judged by themselves, but I would argue that the logical structure of shame and guilt are the same. Thus, to the extent that reality participants are shameless, they are also guilt free.

References

Andrejevic, Mark. 2004. *Reality TV: The Work of Being Watched.* New York: Rowman & Littlefield.

Appadurai, Arjun. 1996. 'Disjuncture and Difference in the Global Cultural Economy', in Arujn Appadurai, *Modernity at Large: Cultural Dimensions of Globalization*, pp. 27–47. Minneapolis: University of Minnesota Press.

Bannerjee, Indrajit. 2002. 'The Locals Strike Back? Media Globalization and Localization in the New Asian Television Landscape', *International Communication Gazette*, 64(6): 517–35.

Barrer, Peter. 2010. 'National Hysteria: The First Year of Reality Television in Slovakia', *Journal of European Popular Culture*, 1(1): 7–23.

Barthes. Roland. 1977. 'The Death of the Author', in Stephen Heath (ed. and trans.), *Image/Music/Text*, pp. 142–8. New York: Hill and Wang.

Beck, Ulrich. 2002. 'The Cosmopolitan Society and its Enemies', *Theory, Culture & Society*, 19(1–2): 17–44.

Bielby, Denise D. and C. Lee Harrington. 2008. *Global TV: Exporting Television and Culture in the World Market.* New York and London: New York University Press.

Bignell, Jonathan. 2006. *Big Brother: Reality TV in the Twenty-first Century.* London: Palgrave Macmillan.

Boyd-Barrett, Oliver. 1998. 'Media Imperialism Reformulated', in Daya Kishan Thussu (ed.), *Electronic Empires: Global Media and Local Resistance*, pp. 157–76. London: Arnold.

Chatterji, Shoma. 2008. 'Women on Television: Looking Back on Hum Log', *India Together*, 26 July 2008, available at http://www.indiato-gether.org/2008/jul/wom-humlog.htm (accessed on 21 June 2011).

Cunningham, Stuart, Elizabeth Jacka, and John Sinclair. 1998. 'Global and Regional Dynamics of International Television Flows', in Daya Kishan Thussu (ed.), *Electronic Empires: Global Media and Local Resistance*, pp. 177–92. London: Arnold.

Fiske, John. 1989. *Understanding Popular Culture*. Boston: Unwin Hyman.

Foucault, Michel. 1984. 'What is an Author?', in Paul Rabinow (ed.), *The Foucault Reader*, pp. 101–20. New York: Random House.

Gopal, Sangita and Sujata Moorti (eds). 2008. *Global Bollywood: The Travels of Hindi Song and Dance*. Minneapolis: University of Minnesota Press.

Hall, Stuart. 1980. 'Encoding/Decoding' in Stuart Hall, Dorothy Hobson, Andrew Lowe, and Paul Willis (eds) *Culture, Media, Language*, pp. 128–38. London: Hutchinson.

Harvey, David. 1990. *The Condition of Postmodernity*. Cambridge, MA: Basil Blackwell.

Indiantelevision.com. 2011. 'Bigg Boss 4 Ends on a High Note with 6.7 TVR', 12 January, available at http://www.indiantelevision.com/mam/headlines/y2k11/jan/janmam51.php (accessed on 12 July 2011).

Keane, Michael and Albert Moran. 2008. 'Television's New Engines', *Television & New Media*, 9(2): 155–69.

King, Barry. 2009. '*Idol* in a Small Country: New Zealand Idol as the Commoditization of Cosmopolitan Intimacy', in Albert Moran (ed.), *TV Formats Worldwide: Localizing Global Program*, pp. 273–89. Bristol, UK, and Chicago: Intellect.

Kraidy, Marwan M. 2005. *Hybridity, or the Cultural Logic of Globalization*. Philadelphia: Temple University Press.

———. 2009. 'Rethinking the Local–Global Nexus through Multiple Modernities: The Case of Arab Reality Television', in Albert Moran (ed.), *TV Formats Worldwide: Localizing Global Programs*, pp. 29–38. Bristol, UK, and Chicago: Intellect.

Magder, Ted. 2009. 'Television 2.0', in Susan Murray and Laurie Ouellette (eds), *Reality TV: Remaking Television Culture*, 2nd edition, pp. 141–64. New York and London: New York University Press.

Manchanda, Usha. 1998. 'Invasion from the Skies: The Impact of Foreign Television on India', *Australian Studies in Journalism*, 7: 136–63.

Mathique, Adrian Abbott. 2009. 'From Monopoly to Polyphony: India in the Era of Television', in Graeme Turner and Jinna Tay (eds),

Television Studies after Television, pp. 159–67. London and New York: Routledge.

McChesney, Robert W. 2001. 'Global Media, Neoliberalism, and Imperialism', *Monthly Review*, 52(10): 1–19.

Moran, Albert. 2006. *Understanding the Global TV Format*. Bristol, UK and Portland, OR: Intellect Books.

Morley, David. 1980. *The "Nationwide Audience"*. London: BFI.

Ouellette, Laurie and James Hay. 2008. *Better Living through Reality TV*. Malden, MA: Blackwell.

Robbins, Kevin and Aku Askoy. 2005. 'Whoever Looks Always Finds: Transnational Viewing and Knowledge-Experience', in Jean K. Chalaby (ed.), *Transnational Television Worldwide: Towards a New Media Order*, pp. 14–42. London and New York: I.B. Taurus.

Robertson, Roland and Kathleen White. 2007. 'What is Globalization?', in George Ritzer (ed.), *The Blackwell Companion to Globalization*, pp. 54–66. Malden, MA: Blackwell.

Roscoe, Jane. 2004. '*BIG BROTHER* AUSTRALIA: Performing the "Real" Twenty-four-seven', in Robert C. Allen and Annette Hill (eds), *The Television Studies Reader*, pp. 311–21. London and New York: Routledge.

Schiller, Herb. 1971. *Mass Communications and American Empire*. Boston, MA: Beacon Press.

Sender, Katherine. 2011. 'Real Worlds: Migrating Genres, Travelling Participants, Shifting Theories', in Marwan M. Kraidy and Katherine Sender (eds), *The Politics of Reality Television: Global Perspectives*, pp. 1–11. London and New York: Routledge.

The Telegraph. 2011. 'Sallu & Sanju in *Bigg Boss*', 21 July.

Thomas, Amos Owen. 2005. *Imagi-Nations and Borderless Television: Media, Culture and Politics Across Asia*. New Delhi, Thousand Oaks, and London: Sage Publications.

Thomas, Amos Owen and Keval J. Kumar. 2009. 'Copied from Without and Cloned from Within: India in the Global Television Format Business', in Albert Moran and Michael Keene (eds), *Television Across Asia: Television Industries, Programme Formats and Globalization*, pp. 122–37. London and New York: Routledge.

Thussu, Daya Kishan. 1998. 'Localizing the Global: Zee TV in India', in Daya Kishan Thussu (ed.), *Electronic Empires: Global Media and Local Resistance*, pp. 273–94. London: Arnold.

Timmons, Heather. 2011. 'In India, Reality TV Catches on, with Some Qualms', *The New York Times*, 11 January, available at http://www.nytimes.com/2011/01/10/business/media/10reality. html?_r=0 (accessed on 7 October 2013).

Tomlinson, John. 1997. 'Cultural Globalization and Cultural Imperialism', in A. Mohammadi (ed.), *International Communication and Globalization*, pp. 170–90. London: Sage Publications.

Wahab, Juliana Abdul. 2010. 'Malaysian Reality TV: Between Myth and Reality', *Malaysian Journal of Communication*, 26(2): 17–32.

Waisbord, Silvio and Sonai Jaslin. 2009. 'Imagining the National: Television Gatekeepers and the Adaptation of Global Franchises in Argentina', in Albert Moran (ed.), *TV Formats Worldwide: Localizing Global Programs*, pp. 57–74. Bristol, UK, and Chicago: Intellect.

10

The Saffron Hues of Gender and Agency on Indian Television

SANTANU CHAKRABARTI

Observers of American politics, especially liberal feminists, were caught in a bind when the American Vice Presidential candidate, Sarah Palin, began to couch her statements in the language of feminism. Should Palin's ascent be celebrated, they wondered, despite her opposition to reproductive control, sex education, and many other hard-fought feminist victories (see Traister 2011)? In other words, is the unprecedented prominence of a specific woman in a hitherto patriarchal or anti-feminist milieu in itself a feminist victory? This chapter deals with a similar question in the context of television soaps from India, a country where questions about the relationship of liberal feminism with the right-wing fundamentalist version of women's empowerment were being discussed as far back as two decades ago.

The three soaps that are the object of my study are landmarks in Indian television. They aired on Rupert Murdoch's STAR Plus, and their dominance in terms of television ratings and revenue was unprecedented for Indian[1] television. These soaps, all from the production house Balaji Telefilms, are the so-called K-serials: *Kyunki*

Saas Bhi Kabhi Bahu Thi (Because the mother-in-law was once a daughter-in-law too, STAR Plus, 2000–8), henceforth *Kyunki*; *Kahaani Ghar Ghar Ki* (The tale of every home, STAR Plus, 2000–8), henceforth *Kahaani*; and *Kasautii Zindagii Kayy* (The challenges of life, STAR Plus, 2001–8), henceforth, *Kasautii*. Viewers, it seemed, just could not get enough of these serials.

The success of these K-serials was remarkable by Indian (and global) television standards of viewership. It was almost as if STAR Plus was broadcasting a show with the reach of *American Idol* three times a day, four to five times a week. If Jeff Zucker, Chief Executive Officer (CEO) of NBC Universal, had known about the K-serials (and been a little less ethnocentric), he would perhaps not so easily have labelled *American Idol* 'the most impactful show in the history of television'.[2] On the back of these shows, STAR Plus became the overwhelmingly dominant Hindi television channel. It remained an unchallenged leader in the television market in this period, delivering every week at least 45 of the top 50 shows and sometimes, all 50 out of the top 50 shows (Krishna 2004). The locus of interest for advertisers, viewers, and entertainment journalists shifted quite decisively to STAR Plus, even if private cable and satellite channels (C&S) as a whole continued to trail the state broadcaster, Doordarshan, in terms of reach. These serials were not just commercially successful, they were culturally significant as well. As television scholar, Shoma Munshi (2010: 1–2), described recently:

> Tulsi of *Kyunki* and Parvati of *Kahaani* became the ideal wives and bahus.[3] They, along with the negative women characters, set fashion trends in saris, blouses and jewellery. Even the men got their share of fan following. When Mihir, Tulsi's husband in *Kyunki*, was killed, women took to the streets in protest.

Munshi actually goes even further than this. She believes that these soaps have 'done women, even in rural India, a great deal of good' (Munshi 2010: 201). But Munshi's laudatory reading of the soaps runs counter to that of commentators and critics writing at the time the shows were first launched. At that time, pejorative terms such as *saas–bahu* soaps[4] or K-serials[5] were coined to refer to these three shows and others of their ilk that were rapidly spreading across all channels. For practically the entire lifespan of the soaps—*Kyunki, Kahaani*, and *Kasautii*, all ended in 2008— the adjective used most often to describe them, especially in the English-language media, was 'regressive'. The attitude towards these soaps ranged from mild embarrassment and grudging admiration to patronizing put-downs and outright hostility. Shobhaa De, one of India's most popular English-language commentators, was scathing about the K-serials in a 2003 interview, contending that 'the protagonists' mindsets were 50 years behind the times [and] the "saas-bahu" themes [were] extremely insulting and degrading' (Indiantelevision.com 2003). Television critic Poonam Saxena was almost celebratory in 2009 about the dethroning of STAR Plus from the number one position after nine long years. The K-serials, she wrote, were typified by 'over-madeup scheming vamps, the multiple marriages and extra-marital affairs, the scripting gimmicks (amnesia, time jumps etc), and the crude special effects (jarring zoom-ins and zoom-outs)' (Saxena 2009).

Activists and academic researchers were equally critical. A 2001 report by the non-governmental Centre for Advocacy and Research (CFAR) expressed concern at the negative portrayal of women on these shows. It criticized the increased depiction of emotional violence, especially the portrayal of women as aggressors against other women, and the distortion of men–women relationships (Shivdasani 2001). In a 2003 paper in the well-known journal, *Economic and Political Weekly*, the CFAR argued that there was in these serials, a 'fairly rigid gender characterization along the expected stereotypes of women and men' (CFAR 2003: 1687).

Of late, though, some of the academic viewpoints seem to have shifted. A longitudinal panel data study from the American National Bureau of Economic Research (NBER) suggested that the positive effects of cable television on issues such as female school enrolment and decreased son preference are large, 'equivalent in some cases to about five years of education…and move gender attitudes of individuals in rural areas much closer to those in urban areas' (Jensen and Oster 2007).[6] Another scholar, Ipsita Chanda (2005), has found evidence to indicate that women often feel empowered by these soaps. Munshi (2010: 192) has criticized the 2001 CFAR report saying that 'the aim of the study presuppose[d] the results'. She argues that these soaps should rather be viewed as sites of contestation and commends them for their portrayal of 'strong women and real issues' (Munshi 2010: 181). Even the writer Shobhaa De has come around in recent years, approvingly noting the fervent appeal that these serials had 'everywhere', even in Pakistan (Munshi 2010: 190). It is in the contradiction between these two sets of responses—the earlier, largely condemnatory, and the later, largely laudatory—that I locate the present study. My aim is to understand why this contradiction emerges, and if possible, to resolve it.

To do that, I first show how the K-serials were a definitive break at that specific point in time from the ones that had immediately preceded them, especially in placing their stories in the milieu of the urban, affluent, extended family and in bringing to the fore the bahu of that extended family. I then argue that the apparently contradictory responses can be reconciled (if only partially) by understanding the context in which this rather sudden shift took place—a task that has not yet been undertaken with any degree of thoroughness. I also show how strong a debt the K-serials owe to the ideological constructions of the woman and the family by the Hindu right in India. I trace the complex place of gender in the discourse of the Hindu right, showing therefore that the representation of gender and agency on these shows is tinged with a saffron (the colour most associated with the Hindu right) hue. In so doing, I argue that televisual representation cannot be studied without taking into account the historical trajectory of such representation on that medium, nor can it be studied without coming to grips with the sociological condition that it is purporting to represent. And, I conclude, it is not enough for observers of television to celebrate agency without taking into account the field (both televisual and social) in which such agency seems to work. I show therefore that the consequences of celebrating agency and empowerment in a right-wing milieu necessarily limit itself to the celebration of very specific kind of female agency. In other words, the answer to Katie Couric's question, 'Can you be a conservative feminist?' (Traister 2011: 275), must be 'not really'!

This chapter is derived from a larger study of the K-serials that looks at the political, economic, and socio-cultural contexts in which these shows were launched to great success. I base my comments here on an analysis of 200 episodes across these three shows. This includes the first 50 episodes each of *Kyunki* and *Kahaani*, the first 15 of *Kasautii*, and significant milestone episodes such as the 100th and the 500th of each. The period of study is from the year 2000 (when *Kyunki* and *Kahaani* started) to 2006, when their relative decline began, but is heavily concentrated towards the earlier part of the decade. This is a conscious choice as I am most interested in unearthing the forces that were in play *around the time of creation* of these shows and in tracing the ideological influences that were inscribed on these shows at birth. This is supplemented with analysis of relevant secondary material such as print interviews with creators and producers of these shows as well as reviews and reports about these shows published in mainline newspapers and magazines, and industry publications.

The K-Serials on STAR Plus: A Decisive Break

The K-serials or saas–bahu soaps were a decisive break from past programming on Indian television, both when compared to earlier shows on state broadcaster, Doordarshan, and when compared to shows of the early years of transnational satellite television from 1992 to 2000. The soaps were conceived and produced by Ekta Kapoor, the driving force behind the family-run production house, Balaji Telefilms. Three of her shows, *Kyunki, Kahaani,* and *Kasautii,* dominated and typified this genre. All of these shows aired on prime time and were broadcast Monday–Thursday initially, and Monday–Friday later on. And because Indian television does not operate seasonally like British or American television, these soaps aired continuously without a break for the entire seven to eight years of their lifespan, thereby magnifying their reach and influence.

These soaps shared characteristic and distinctive aesthetic, visual, and storytelling features. In her book, *Prime Time Soap Operas on Indian Television,* the most comprehensive survey till date of the production aspects of soap operas in the era of STAR Plus dominance, Munshi (2010) identifies some of these features.

She observes that while some features were common to soaps all around the world, there were others which were local innovations. Features like opulent sets, decor, costume, jewellery and make-up, and the dominance of a central female protagonist in driving the narrative, common to most soaps worldwide, were present in the K-serials as well. Common also was the ending of a week's last episode on a cliffhanger and starting a new episode with a 'recap' of what had gone before. Creaky storytelling devices such as a character's return from the dead were also quite widely used (Munshi 2010).

What was unique to Indian soaps, though, was the regular use of plastic surgery and amnesia as explicatory plot devices, sometimes to explain the return of a character presumed dead or sometimes to explain the appearance of a different actor playing the same character. Often, there were 'precaps' highlighting the contents of the next day's episode and there was the notable local innovation, the 'generation leap', where the focus shifted to the younger generations of the family at the centre of the plot, while the most popular characters were aged by decades (though they looked scarcely older, apart from a few streaks of made-up grey in the hair). And there was the use of the 'swish pan shot', where the same action was shown three times in quick succession, accompanied by loud music, with the characters or scenes often depicted in different colours each time (Munshi 2010). These shows also brought to the fore a very specific representation of the family and the woman, but to properly situate these representations, we need to take a quick look at the kind of representations of women seen on Indian television in the years immediately preceding the K-serials.

The Woman on TV: Before the K-Serials

It is understandable that some present-day scholars believe that the K-serials have been unfairly critiqued even though they depicted 'strong women and real issues'. But it needs to be kept in mind that earlier critics who labelled the K-serials as 'regressive' were responding to the earlier serials, both on state-run Doordarshan before the advent of private television in 1991 and in the earlier days of private C&S television, in the period 1991–9.

Hum Log (1984–5, Doordarshan) and *Buniyaad* (1986–7, Door-darshan), two of the earliest successes of Indian television, presented women largely within the environment of the home. The mythological *Ramayana*, still perhaps the most watched Indian television series of all time, had at its heart the travails of Sita, the embodiment of the respectful, suffering, sacrificing Indian woman. As Purnima Mankekar (1999: 10) has argued, while the trope of the *Bharatiya nari* or the ideal Indian woman pre-dates television, 'Doordarshan occupied a central place in constituting female viewers not just as women but as *Indian* women' (emphasis in original).

Despite their many faults, Doordarshan shows did portray many women protagonists dealing with real-life issues outside of the family home. The character Renu Verma, on the classic comedy series *Yeh Jo Hai Zindagi* (This life that is, 1984), was depicted doing a job, her skirts and dresses superficially indicating her 'modernity', and her spirited retorts to her husband indicating a level of equality. Even more engaged with the world outside the home was the homemaker–activist, Rajani, crusading against corruption in *Rajani* (1985). Priya Tendulkar, the actress who played the role, became one of the early stars of Indian television and her eponymous character became a culturally significant metaphor for urban middle-class angst against institutionalized corruption in India. Five years later, another strong female character again struck a significant chord with viewers: the policewoman Kalyani Singh on *Udaan* (Flight, 1990), played by the serial's writer, Kavita Choudhary. As Mankekar (1999: 139–49) puts it, *Udaan* was 'exceptional in its construction of a complex and introspective protagonist...whose observations on corruption in the police force coexist[ed] with the pleasure she derive[d] from her own ascent to power', even if the critique of the patriarchal family structure was softened by the creator herself and did not go 'far enough in its advocacy of feminist activism'. Abhijit Roy (2008: 38) sums up the scenario rather well: 'Many of these serials telecast on the much-demonized Public Television seem to have been sensitive at least in their figuration of certain realms of life about which the soaps in the era of satellite television are terribly amnesiac'.

Women were not restricted to the home and the hearth between 1991 and 1999 either, the early years of the C&S era. The programming on private channels was directed at a supposedly urbanized, globalized, middle-class audience, depicting topics like adultery, rape, sexual harassment, working women, and extra-marital relationships, mostly unknown in the monopoly days of Doordarshan. As Page and Crawley (2001) show, they focused largely on the 'new bold woman'. They contend that these shows offered a 'variety of new role models to the urban middle class...provoked much controversy in the process' (Page and Crawley 2001: 166). The most prominent family structure in these shows was the 'aspiring nuclear families' (Page and Crawley 2001: 155) or extended nuclear families (that is, nuclear families plus the paternal grandparents). These shows often featured the woman walking out on the marriage or refusing to graciously welcome back the straying husband as she might have been expected to do as a traditional Indian woman. Tara in *Tara* (1993–7), Saavi in *Hasratein* (Desires, 1996–7), Priya in *Saans* (Breath, 1998–9), and Pooja in *Kora Kagaz* (Burnt paper, 1998–9) were all strong women characters in their own right, even if the serials themselves were often wracked by anxiety surrounding the purported decay of the institution of marriage. The failure of such serials to see their storylines through to their logical conclusions was a forthright indictment of the institution of marriage itself.[7]

Clearly, Indian television had seen more than a few strong, independent women, especially in the Doordarshan era. By itself, then, the fact that K-serials depicted women with strength of character at the centre of action was not by any means a remarkable departure from the past. What was remarkable was that these women protagonists (Tulsi from *Kyunki*, Parvati from *Kahaani*, and Prerna from *Kasautii*) were now completely ensconced in the home and within the extended joint family. As the CFAR (2003) paper pointed out, the women of the K-serials were overwhelmingly confined inside their homes, rarely venturing out into work or even into public spaces. In contrast, Renu, Rajani, Kalyani, and Shanti in the earlier Doordarshan shows were often successful in the outside world and their identities were constituted not merely by their status in the family unit alone.

Nor were the characters in the shows of the early C&S era bound as rigidly by the strictures of tradition or the walls of the family home as much as the heroines of the K-serials. At the superficial level, viewers could see Tara smoke and drink without apology (and managed to shock a legion of puritanical television viewers); Saavi leave her husband to live in with her lover; and Priya and Pooja finding their own identities distinct from their philandering husbands. While one can argue that these are indeed not more than severely attenuated indicators of agency, Purnima Mankekar has suggested that even these depictions were quite radical for Indian popular culture, especially considering the fact that the traditional Indian family continued to be central to the narrative of contemporaneous Bollywood film. She suggests that the emergence of the supposedly traditional Indian family form in the K-serials constituted a backlash to these depictions of desire and independence in the Indian woman (Mankekar 2004). (As we will see later, that tells only a part of the story.)

K-Serials: The Return of the 'Traditional' Indian Housewife

With the advent of the K-serials in 2000, the extended joint family and the issues of property and relationships around it became the central concern of television. The protagonist was now ensconced within the walls of the family mansion. The family became central not only to the narratives of the shows but also in marketing and promotional efforts that STAR Plus carried out to push these shows. A key marketing innovation was the creation of a televised awards show where the winners would be the characters and *not the actors playing these characters* (Unnikrishnan 2003). The show was named *The STAR Parivaar Awards* (The STAR Family Awards) and the award categories were Favourite Son, Favourite Daughter-in-Law, Favourite Mother-in-Law, Favourite Paternal Grandfather, and so on. The communication and theme song for these awards spelt out the family thematic clearly, proclaiming that the viewer and the characters on STAR Plus soaps belonged to one undivided family. 'We had come as guests once,' the lyrics went, 'Little did we know that we would become one of your family. And the relationships we depicted on screen would be formed with you as well.' In short, the traditional family occupied pride of place in the environment of the K-serials and the woman's place was now unambiguously the home.[8]

Munshi (2010: 217–18) contends, however, that 'the unflinching, uncompromising capacity to suffer endlessly and follow the right moral path...even when faced with familial displeasure— permits soap heroines [of the K-serials] to assume a strong and powerful position that, in fact, questions patriarchal authority'.[9] Without entirely disagreeing with this contention, we should also ask if the depiction of a policewoman triumphing at work (say, Kavita), or a woman social activist (say, Rajani), or a woman walking out on her marriage to start a live-in relationship (say, Saavi) does not question patriarchal authority even more strongly. So why is it that the bahu *identity* becomes suddenly central to representations of women on television? And equally, why does the supposedly traditional Indian family become so central to television serials at the end of the twentieth century? And why can the heroine of the K-serials now question patriarchy only from within the highly patriarchal extended joint family, even though the heroines of earlier shows were often depicted outside of the home and/or outside the bounds of the traditional joint family?[10]

The K-serials and Hindutva[11]

Munshi notes the almost overwhelming presence of Hindu rituals, iconography, and representation in the K-serials and acknowledges that they conflate an Indian identity with a Hindu identity. But after receiving assurances from her respondents (the producers of these soaps) that 'there is no deliberate attempt... for this predominance of Hindu identities', Munshi (2010: 179) does not problematize this any further. I believe, though, that this can be one of the most fruitful avenues for investigation into the production and success of these shows. As I see it, this lack of problematization arises partly from an insufficient appreciation that the representations of women on the K-serial were a significant departure from depictions of women in earlier eras of television, especially the early C&S era. The majority of families on Doordarshan shows, as on early C&S shows, were urban, upper-caste Hindu families, as were the ones on the K-serials. So, a natural tendency might be to look at Tulsi and Parvati and say, what's new? That is understandable but not complete. The lack of problematization, though, has also to do with the fact that observers have not detected expressions of the most virulent, vitriolic kind of Hindu nationalism that came to a boil in India in the early 1990s, even if they have found some of these shows complicit in peddling a Hindutva agenda (see, for example, Deshpande 2009: 20 and Roy 2011: 22–4). That is, none of these shows engaged in active and easily discernible minority bashing; and in none of them would you find patriarchal expressions at their most stomach-churning form, both of which were a staple of the discourse of the Rashtriya Swayamsevak Sangh (National Self-Volunteers Organization, henceforth RSS), the fount organization of the organized Hindu right.

This, however, misses two very salient points. One, Hindu nationalism has always viewed the creation of the Hindu nation as a task of primary importance even preceding the creation of the Hindu state (see Blom Hansen 1999; Jaffrelot 1996). The cultural project, therefore, might draw strength from the success of the political project but even when the political project falters (as it did in the national elections of 2004 and 2009), the RSS marches ahead with its cultural project. Two, and even more importantly for our purposes, commentators who celebrate the representations of women on the K-serials ignore the fact that modern-day right-wing fundamentalisms are not necessarily inimical to the expressions of women's agency. The discourses of the Hindu right have not been shy of fostering and drawing on expressions of women's strength, but the key question has always been the delineation of the arena in which this strength is to be exhibited and the extent of that strength. To understand why the ideologies of the K-serials (despite all their expressions of female energy) don't run counter to the ideologies of the Hindu right, we need to engage with the discourses of the Rashtra Sevika Samiti (National Women Volunteers Committee, henceforth Samiti) the women's wing of the RSS.[12] As we shall see, the discourses of the Samiti have run parallel to and have often overlapped with (but are not identical to) the mainline discourse of the RSS. Any understanding of the representation of women in the K-serials has to first engage with the alternative discourse of women and gender as purveyed by the Samiti as well as the way in which these were articulated and practised in the 1990s by the Hindu right.

Gender, Agency, and the Hindu Right

Historian Tanika Sarkar made an important point in 1991 when discussing the first efflorescence of right-wing Hindu women's participation in the public sphere, especially in militant activities: 'In fact the new Hindu woman citizen is cast in a mould that is very close to bourgeois feminism. There is no denying that it [that is, the discourse of the Samiti] does empower a specific and socially crucial group of middle class women, if not in an absolute feminist direction then definitely in a relative sense' (Sarkar 1991: 2062). In an important collection of essays on women and the Hindu right, Amrita Basu (1995: 171) pointed out that the 'gendered imagery and the actual roles of women in Hindu nationalism are far more diverse than in [standard] fundamentalist depictions'. The discourse of Hindu nationalism, then, does not actively discourage women's education and work. But it certainly expects primacy to be given to motherhood, because it is with the mothers that the critically important task of reproducing the cultural values of Hindu culture, and therefore the establishment of the Hindu nation, rests (Blom Hansen 1994). If Hindu nationalism has always given primacy to the creation of the Hindu nation, the Samiti believes that the institution upon which the nation is based is that of the Hindu family. If the Hindu nation has been weakened and is in need of rediscovering the fundamental principles of Hindutva, it has largely been because this ideal family has been destroyed by intermarriage; and this has gone hand in hand with indignities heaped upon Hindu women by Muslim men, without any equivalent retaliation from weak and effeminate Hindu men (see Agarwal 1995; Bacchetta 2004).

In the late 1980s and into the 1990s, as militant Hindu nationalism was going from strength to strength, more and more women were being engaged in Hindu nationalist activity. Women were actively encouraged to transgress boundaries of the home and even of geography, as long as they were doing so in the service of Hindu nationalism. As Tanika Sarkar and Urvashi Butalia (1995: 3) describe, this posed a serious challenge to feminists: 'Politically and methodologically this assertive participation of women in right wing campaigns pulled many of our assumptions into a state of crisis for we had always seen women as victims of violence rather than its perpetrators and we have always perceived their public, political activity and interest as a positive, liberating force'. Some of the most violently anti-Muslim rhetoric was spouted by women leaders of the movement such as Uma Bharati and Sadhvi Rithambhara (who became infamous for her obscenity-laden, vitriolic declamations of anti-Muslim rhetoric). As Paola Bacchetta (2004: 2) has conclusively demonstrated, the activism of Hindu nationalist women in this period was supported by a 'specifically feminine Hindu nationalist discourse', exemplified best in the discourses of the Samiti.

So, where is the problem, one might ask, even if female empowerment comes couched in the Hindu nationalist shade of saffron?[13] The problem is twofold. One, even while certain female leaders of the Hindu right have themselves grappled with gender inequality, it has always been restricted to the cause of Hindu women. Otherwise, '[a]t their most benign, they render Muslim women invisible; more often they seek to annihilate Muslim women' (Basu 1995: 164). That is, while women's self-constitution as an active political subject was certainly enabled, it was done in 'dangerously unprecedented ways' (Sarkar 1991: 2057). The other equally significant problem is that the empowerment of women driven by the doctrines of right-wing Hindu nationalisms (and for that matter, right-wing doctrines in general) is ultimately severely constricted. Both of these points were illustrated perfectly by the developments at the end of the 1990s.

Bottling the Genie: Hindutva and Women's Energies

As we just saw, the strategic deployment of women's agency helped make Hindu nationalism a potent force in the 1980s and the 1990s. But as Thomas Blom Hansen (1994) has argued, the Hindu right's response to greater visibility of women in the public sphere was to deploy a strategy of 'controlled emancipation'. He argues that there were two Hindu nationalist strategies adopted with respect to women: one asserted the primacy of motherhood, while the other tried to 'suture gender conflicts through the controlled emancipation of women under the protective canopy of Hindu nationalist organisation' (Blom Hansen 1994: 82). By the middle of the 1990s though, women were being told to retreat back into the home and were no longer as visible on the frontlines of the Hindu nationalist movement. When Tanika Sarkar went back in April 1999 to talk to the respondents she had interviewed in the early 1990s, she found their circumstances and their outlook significantly changed. Specifically, she found that (RSS ideologue) 'Golwalkar's restricted and restrictive strictures on domesticity and the homebound women have been retrieved and refurbished' (Sarkar 1999: 2162). Earlier, representatives of the Samiti were very clear that the notion of sacrifice was bound up with active fighting. Samiti publications argued that a woman who was able to defend herself acquired greater respect in society. The Samiti magazine, *Jagriti*, argued in 1991 that it was extremely important for women to be economically independent for them to achieve comprehensive development. As Sarkar argues, 'the new Hindu woman [was] cast in a mould which is very close to bourgeois feminism' (Sarkar 1991: 2062).

Yet, by the end of the 1990s, representatives of the Samiti were saying that too much insistence on women's rights was leading to the dissolution of the incubators of the Hindu nation—the family. In the words of a prominent Samiti activist, 'they [feminists] teach women about their rights, they tell them to fight their men about these rights. We teach them how to sacrifice themselves to keep the family together' (quoted in Marsh and Brasted 2007: 296–7). Women's subjectivities were now again being anchored in their primary roles as wife and mother and the older trope of women's 'strength in sacrifice' was being excavated again. In parallel, Sarkar (1999: 2165) suggested, 'The new consumerist self-absorptions of the middle class woman, fanned by the ad-culture and the flood of beauty-aids, cosmetics and household-gadgets, are encouraged, since they provide the economic survival of much of the country's manufacturing–trading classes. And this class is also the major basis for the political support of the Sangh parivar.'

The retreat into the family—not to say there was ever a total rupture from it—had another facet. In addition to drawing strength from tradition in the face of the encroachments of globalization, it was also a bulwark against the Hindu right's most vilified punching bag—the Muslims.

Hindutva and the Politics of the Family

The continued success of the Hindu nationalist project in the 1980s and 1990s meant that the bridges between different religious communities were often being destroyed, especially those between Hindus and Muslims. Social exclusion was continually increasing and Muslims were falling economically behind as well, unable to get even the crumbs from a new globalized economic regime (see, for example, Alam 2010 and Shariff and Razzack 2006). Despite this, Hindu nationalists continued to argue that state favouritism towards Muslims was the reason why disadvantaged Hindus were losing out on opportunities. Muslims continued to be demonized as the threatening other (Jaffrelot 1996).

The older leadership of the Samiti in the 1990s was increasingly uncomfortable with the emergence of a younger generation of women, who, instead of devoting time to building the ideal Hindu family (the microcosm of the Hindu nation, as we have seen), pursued careers or education, earning the disapproval of the leadership for fostering the 'mummy and daddy culture of nuclear families' (Blom Hansen 1994: 88). Ironically enough, there is not enough sociological evidence to say with certainty that the joint family, so beloved of Hindu nationalists and referred to even by academics as the 'traditional' family, had actually been the dominant family form in India in years past. The perception that the joint family is the traditional Indian family form and the notion that the form has increasingly been at threat due to the advent of modernity was largely a contribution of British Orientalists and their successors, the Indologists.[14] If the joint family form was not the defining form of the family for most Indians, then, obviously, the theory of its disintegration was highly contestable. However, it was possibly the defining family type for a particular type of Indian: north Indian, urban, upper-caste traders, clerks, and business folk, which formed the core support base for the Sangh. But A.M. Shah has presented a wealth of historical and empirical data to suggest that there was no significant dissolution of the joint family structure post the advent of industrial capitalism in India, and if at all it has been happening, it has been happening for professional and Westernized upper middle classes (Uberoi 2000). Still, as Patricia Uberoi (2000: 130) rues, 'the public at large, and even many social scientists, remain addicted to their misconceptions'.

True or not, the fear of the traditional family splitting apart was central to the Hindu right's understanding of the changes wrought by globalization and economic liberalization in the late twentieth century. This misconceived idealization of the nature of the Hindu family was counterpoised against the even more egregiously pernicious misconceptions of the Muslim family, and increasingly made commonsensical during the ascendance of the Hindu right in the last two decades of the twentieth century. Even as attempts were made to combat the perceived nuclearization of the Hindu family under the cultural invasion from the West and establish the Hindu joint family as the repository of Indian tradition, tropes about Muslim fertility and family sizes were deployed extensively. The Hindu right argued (and continues to argue even today, if a little more *sotto voce*) that Indian Muslims gained an unfair demographic advantage from the existence of separate family laws for the adherents of different religions given that Muslim religious law allows all Muslim men to marry four wives. The Muslim family is typified in the vitriolic formulation, '*Hum paanch, hamare pachis*' ('We five, our twenty-five'), a spin on the Indian official family planning mantra, 'Hum do, hamaare do' ('We two, our two'). This formulation not only indicates the supposed lack of Muslim interest in family planning (Anand 2007), but by extension, it highlights their lack of patriotism in carrying out the national project of population control. It articulates, therefore, the pervasive anxiety that prevails in Hindu nationalist circles about the Muslim 'plot' to overrun Hindus demographically by producing more children, which will apparently result in there being more Muslims than Hindus in India by 2051 (Reddy 2002).

Almost needless to say, these discourses prevailed, even though, in reality, there is less polygamy among Muslims than among other groups, such as tribals, and rates of Muslim polygamy are comparable to rates of Hindu polygamy (Puniyani 2003). In addition, Muslim family law itself underwent significant changes since the 1970s, giving earlier wives the right to divorce their husband if he practised polygamy. Further proving the empirical emptiness of these discourses is the fact that the growth rate differential between Hindus and Muslims is marginal at best, and family planning has more to do with socio-economic factors rather than religious ones (see, for example, Shariff and Razzack 2006 and Subramanian 2008).

Despite all such evidence to the contrary, the nature of the Muslim family has been positioned as distinctively different from the Hindu one. The representation of the family on television itself is therefore, in a sense, a political statement. And the decline (and eventual disappearance) of the nuclear family and the resurgence of the joint family on Indian television took place then in an environment where the joint family was being celebrated as a bulwark against both the supposed excesses of the emancipated woman and the purportedly rampaging Muslim population growth.

Women on the K-Serials: Tracing the Influences

Kyunki Saas Bhi Kabhi Bahu Thi debuted on 3 July 2000, heralding an era in which affluent, urban, upper-caste Hindu joint families seemed to take over television. Both STAR Plus executives as well as Ekta Kapoor (the creator of and prime mover behind these shows) were sure that what they were presenting was 'India'.

According to Shailja Kejriwal, the creative head of STAR Plus, '[A]n important part of when *Kyunki* began was to show how the new *bahu* of the house, Tulsi, saw that her mother-in-law's generation was not taking good care of their elders' (quoted in Munshi 2010: 119). Note how the new generation of women is now being vested with the responsibility of guiding an earlier generation that seems to have abdicated its responsibility towards upholding 'Indian' tradition. According to Ekta Kapoor, 'the one subject which holds eternal interest for...Indians is the family—every Indian family is bound by traditions, festivals, etc., and every family tends to celebrate occasions with relatives...' (quoted in Munshi 2010: 118). There is no mention of any element of society that is not tied by kinship bonds. And obviously, then, there is no space for religious minorities, given the rare occurrence of inter-religious marriage in India.

While the Hindu right and the BJP (the political arm of the Sangh Parivar) government had been extremely negative about the C&S television soaps of the 1990s (McMillin 2003), they embraced the K-serials rather more warmly. If indeed the K-serials questioned patriarchy as much as Munshi (2010) suggests, it is surprising that it did not ruffle any right-wing feathers, especially given the significant popularity of these shows. On the contrary, an article in the RSS mouthpiece, *Organiser*, called Smriti Irani, the actress who played the protagonist Tulsi on *Kyunki*, 'the ideal daughter-in-law that the television has ever shown' (*sic*) (Nigam 2005: 41). Little wonder then that Irani was made a BJP candidate for the Lok Sabha in the 2004 elections and became, in 2010, a secretary in the national executive team of Nitin Gadkari, the new and more hardline president of the BJP, the political arm of the Hindu right. Given how little Irani was known before taking on the role of Tulsi, the BJP were clearly aligning themselves with the idea that the character Tulsi embodied. But these ideas about the nature of contemporary womanhood were prevalent in the Sangh Parivar *before* the shows launched.

For example, Amrita Basu (1995) quotes the BJP leader and ideologue (then recently appointed President of the party), Murli Manohar Joshi, proclaiming in 1991 that the BJP was 'with women power' in their struggle. But then he goes on to say, 'Please forgive me for saying that you have to change yourselves, for women forget their own sufferings when they become mothers-in-law' (as quoted in Basu 1995: 172). He has no problems with women exerting their agency, but he is equally certain that the agency has to be exerted in the domestic sphere. The very title of the biggest hit on Indian television of the post-Doordarshan era embodies and echoes Joshi's schematic: *Kyunki Saas Bhi Kabhi Bahu Thi*, that is, 'Because the Mother-in-Law was once a Daughter-in-Law too'. And Joshi was saying this a good half a decade before the K-serials were conceived.

The names of the leads (Tulsi from *Kyunki* and Parvati from *Kahaani*) were not accidental or coincidental choices. The tulsi (or holy basil) plant is worshipped by traditional Hindus and is associated with the legend of Krishna, another incarnation of the God Vishnu. Parvati is not only the name of the mother goddess, benign consort of Shiva, but also one of the names by which Bharat Mata—the symbolic space of the motherland worshipped as a goddess—is referred to in Samiti literature. While Parvati in Hindu myth is a benign goddess, in Samiti literature, Bharat Mata–Parvati is variously referred to as 'protector of society' and 'the very source of all power' (Bacchetta 2004). Not only does the Samiti assign agency to the Hindu woman, it goes far enough to redefine the concept of *pativrata* (dedicated to/worshipful of the husband), so that not only the husband but also the nation is her god. A woman's self, then, in the words of Bacchetta (2004: 7–8), 'implies not just the individual self but also family, society, nation, religion and culture [and is thus] relational, and merged in other (always bi-gendered) entities'. Therefore, she suggests, in the unlikely situation of a conflict between duty towards husband and duty towards the nation, it was the latter that would prevail. But while 'women would be free of the *individual* Hindu male [they would] remain bound by the Rule of the *collectivity* of Hindu nationalist males' (Bacchetta 2004: 42; emphasis in original).

The idea that Tulsi and Parvati are 'fighter[s] in the struggles against evil', as Munshi (2010: 193) suggests, obviously begs the question why they do not fight evil in the wider world like Rajani or Kalyani did on Doordarshan way back in the mid-1980s and

Channeling Cultures

early 1990s. And strong as these women on the K-serials might be, even in the early years of the twenty-first century, they are not allowed to walk out on their families or on philandering husbands; nor are they allowed to exist as beings with a sexual identity of their own. This is especially striking when one considers that a whole set of shows aired in the early C&S years, just before the K-serials, 'showed women actively, sometimes aggressively, pursuing erotic pleasure and facing the social and emotional consequences of doing so' (Mankekar 2004: 422). These restrictions on the women of the K-serials underscores, once again, the Samiti version of feminism (if at all we can call it that), which preaches understanding and sacrifice in the interest of preserving the family structure.

In this undying structure, women don't really work unless compelled to by financial pressures. Blom Hansen (1994) tells of hearing from women activists of the Hindu right that while women do need to work for money these days, if the family does not need money, there is no need for women to work. In the worldview of the Hindu right then, women do not work for their own fulfilment or developing subjectivity outside of the home, but they do it out of compulsion. And so in the K-serials, women do not work outside of the home. If they are sometimes called to work, it is usually in the family business and that too because the male workers are absent from the picture (usually due to contrived plot reasons). You only need to compare these to Doordarshan shows like *Udaan*, where the protagonist is a police officer, and or even *Yeh Jo Hai Zindagi*, where Renu works in an office, to know why accusations of being regressive greeted the advent of the K-serials.

But what has escaped the attention of many (if not all) commentators is the complete lack of Muslim characters on the show. There is not one Muslim character of note in any of the K-serials throughout its run, not even in negative or heavily patronized characterizations (as Bollywood movies would regularly depict). Obviously, this is partly explained by the fact that each of these serials operates largely in the domain of the family. But even when peripheral characters are introduced or other families are showcased, none of them are Muslims, or even Christians for that matter. Note, though, that this is not to suggest that there is any evidence of active discrimination towards Muslims as far as the cast and crew are concerned: quite a few of the cast members of these devoutly Hindu households are played by Muslims and there are Muslims on the crew as well. But perhaps the problem lies in the fact that despite the presence of Muslims on the set and behind the scenes, there is no Muslim presence on screen in any of the shows.[15] For that matter, there are no significant Christians, Buddhists, or Sikhs either on any of the 1,833 episodes of *Kyunki*, or the 1,653 episodes of *Kahaani*, or the 1,420 episodes of *Kasautii*. This is in stark contrast with the show that is often credited with being one of the key inspirations behind the K-serials, as also the show that was dominant in the ratings till the K-serials dethroned it: the Zee TV show, *Amanat* (1997). *Amanat* is the story of seven sisters and their father, Lahori Ram, who, as the name indicates, came over from Lahore in Pakistan after India's partition into the two countries of India and Pakistan in 1947. Nonetheless, Lahori Ram's best friend is a Muslim, called Ahmed Chacha (uncle) by his daughters, and appears in a significant number of the episodes and is part of significant plot developments.

The absence of minority characters in the K-serials is also glaring when compared to the shows on state-controlled Doordarshan. These shows were more overtly influenced by the Indian state's official secularism and not only featured minority characters prominently, they were often situated in minority communities and strongly advocated secular tendencies (at least when the state was not trying to play soft Hindutva politics by airing Hindu mythological epics like the *Ramayana*).[16] Sanjay Asthana (2008) has shown how Doordarshan serials like *Gul Gulshan Gulfam* and *Choli Daman* engaged religion and secularism on the same plane, not only featuring Muslim or Sikh characters as co-protagonists but also focusing on inter-community engagement and critiquing the trend of desecularization. For all the 'strong women and real issues' that the K-serials engaged with, not once did they engage with issues of inter-religious equations, or caste, or for that matter, issues of economic inequality, the three major issues that continue to define India even in the second decade of the twenty-first century.

It is not just that the characters were Hindus and no Muslims were seen anywhere close to them; it is that the Hindu-ness portrayed was overt and indivisible from their identities. The characters in the K-serials epitomized the consuming rich but they wore their Hindu identity on their sleeves. The Hindu-ness was visible from the very start of the shows. For example, the opening title montage of *Kahaani* started with a ceremonial lamp being lit and moved on to the shot of a pair of Hindu gods, cutting to that of the family members praying, and finishing finally with another shot of the gods.[17] Plot points of significant episodes revolved around crises of ritual and religion in a manner that was never the case for earlier shows on Indian television. For example, an episode in which characters from one show 'crossed over' to another— in another interesting Indian soap opera innovation— revolved around a stolen idol of a Hindu god; Hindu festivals, worshipping, and rituals were seen with monotonous regularity (Munshi 2010). For the ideologues of Hindu nationalism, these rituals are what can primarily help in defining Hindu-ness. In fact, the foundational text of Hindutva—V.D. Savarkar's *Hindutva: Who is a Hindu?* (2003[1942])—establishes common laws and rites, such as religious festivals, as criteria of Hindu-ness (that is, Hindutva).

Once you remember that for the Samiti, the nation is anchored in the Hindu family, it is not too much of a stretch to say therefore that the K-serials were engaged in symbolically erasing minorities from the Hindu nation, just like they were sought to be erased in actuality, a project that was stalled in real life partly due to electoral compulsions (both the exigencies of coalition politics and the BJP's loss of power at the national level in 2004). Be that as it may however, the 'unreal and perverted' women of television shows preceding the K-serials had now been safely contained in the home and within the family. However much agency they expressed there, it would ultimately remain within the four walls of the (admittedly opulent, absolutely fabulous) family home. In other words (and in answer to Katie Couric), a conservative feminism (even if it does exist) is by definition a temporally, spatially, and socially limited feminism. And is a feminism with such boundaries even feminism?

Notes

1. To be absolutely accurate, one should use the descriptor 'Hindi' or 'north Indian' instead of 'Indian' here, given that the channel, Sun TV, was equally dominant in the southern state of Tamil Nadu. But given the numerical strength of the Hindi television audience, such slippage between the terms 'Hindi' and 'Indian' is both unavoidable and not completely inaccurate.

2. Jeff Zucker, CEO of NBC quoted in Bill Carter, 'For Fox's rivals, "American Idol" Remains a "Schoolyard Bully",' *The New York Times*, 20 February 2007, available at http://nyti.ms/16FbMEC (last accessed on 12 January 2012).

3. The Hindi for daughters-in-law, used here to connote the traditional aspects of the role of a daughter-in-law.

4. Because these shows mostly revolved around family politics, often (though not always) revolving around the relationship between the mother-in-law (saas) and the daughter-in-law (bahu).

5. Because the names of the most popular shows produced by Balaji Telefilms started with the letter K (due to numerological beliefs of the progenitor and producer of the serials).

6. Perhaps only in a country where 31 out of 100 children entering the first grade make it past Class X and only nine go on to higher education, can respectable academic studies suggest that cable television influence is equivalent to five years of education. See Yechuri (2009).

7. For a nuanced analysis of the representation of women in the early years of satellite television, see Uma Chakravarti (2000).

8. It must be noted here that women in these serials were occasionally depicted as treading the waters of the 'family business' but that happened so rarely as to be insignificant, and it always took place in the service of a plot point; these characters hardly ever venture into any workplace not associated with the aforementioned family business.

9. Note also that in talking about the positive influence of the K-serials, Munshi draws heavily from the Jensen and Oster (2007) working paper, 'The Power of TV', which is not quite applicable in the context Munshi uses it. First, that working paper focuses on rural women, while the K-serials were available more widely in urban and semi-urban areas. Furthermore, while the working paper only talks about 'cable' without mentioning specific programmes, it is extrapolated by Munshi to mean the K-serials. This is obviously problematic.

10. For our purposes, the joint family is one which is marked by at least two couples (in the case of the K-serials, many more than two) linked by

kinship, living together under one roof marked by commensality, coparcenary financial arrangements, and regular common worship of family gods, with the living unit consisting of at least a common living room, a common entrance, a common dining hall, and (if feasible, but not always) separate bedrooms for each married couple. See Krishna Chakrabortty (2002).

11. The ideology of right-wing Hindu nationalism that seeks to redefine India as a Hindu nation.

12. Incidentally, the first founded of the many affiliated organizations that subscribe to Hindutva ideology and make up what is called the Sangh Parivar (Sangh Family). The best known of this is, of course, the political party Bharatiya Janata Party (BJP), but almost all of the Parivar organizations were founded with a specific objective with each catering to a specified group, for example, workers, diasporic Indians, and so on.

13. Saffron is the color of renunciation in orthodox Hindu thought, often worn by mendicants; and was adopted by Hindu nationalists as the colour that was distinctive to and identified with Hindus.

14. For more on the family size debate in Indian sociology, see, for example, D'Cruz and Bharat (2001); Niranjan *et al.* (2005); Palackal (2007); and Uberoi (2000).

15. This analysis is based on a sample of episodes mentioned earlier. It is hard to conclusively say that there has not been even one Muslim character in the entire run of episodes for each of these shows, but there has certainly not been a Muslim character of primary or even secondary importance across these shows. See, for example, the complete cast pages for cast/character listings for *Kyunki*, *Kahaani*, and *Kasautii*, available at http://www.imdb.com/title/tt0278212/; http://www.imdb.com/title/tt1341970/fullcredits#cast; and http://en.wikipedia.org/wiki/Kasautii_Zindagii_Kay#Cast respectively. The point here is that Muslim characters do not even figure in highly stereotyped or negative forms, as they often did in Bollywood movies—they have been completely written out of the narratives.

16. See Rajagopal (2001) for an account of how the *Ramayana* broadcast was appropriated by the Hindu right in the service of creating what Rajagopal calls 'retail Hindutva'.

17. The video is available at http://www.youtube.com/watch?v=9UKtWddmsuU (accessed on 9 May 2009).

References

Agarwal, Purshottam. 1995. 'Savarkar, Surat and Draupadi: Legitimising Rape as a Political Weapon', in T. Sarkar and U. Butalia (eds), *Women and the Hindu Right: A Collection of Essays.* pp. 29–57. New Delhi: Kali for Women.

Alam, M.S. 2010. 'Social Exclusion of Muslims in India and Deficient Debates about Affirmative Action: Suggestions for a New Approach', *South Asia Research*, 30(1): 43–65.

Anand, D. 2007. 'Anxious Sexualities: Masculinity, Nationalism and Violence', *British Journal of Politics and International Relations*, 9(2): 257–69.

Asthana, S. 2008. 'Religion and Secularism as Embedded Imaginaries: A Study of Indian Television Narratives', *Critical Studies in Media Communication*, 25(3): 304–23.

Bacchetta, Paola. 2004. *Gender in the Hindu Nation: RSS Women as Ideologues.* New Delhi: Women Unlimited.

Basu, Amrita. 1995. 'Feminism Inverted: The Gendered Imagery and Real Women of Hindu Nationalism', in T. Sarkar and U. Butalia (eds), *Women and the Hindu Right: A Collection of Essays*, pp. 158–80. New Delhi: Kali for Women.

Blom Hansen, Thomas. 1994. 'Controlled Emancipation: Women and Hindu Nationalism', *The European Journal of Developmental Research*, 6(2): 82–94.

———. 1999. *The Saffron Wave: Democracy and Hindu Nationalism in Modern India.* Princeton, NJ: Princeton University Press.

Centre for Advocacy and Research (CFAR). 2003. 'Contemporary Woman in Television Fiction: Deconstructing Role of "Commerce" and "Tradition"', *Economic and Political Weekly*, 38(17): 1684–90.

Chakrabortty, Krishna. 2002. *Family in India.* Jaipur and New Delhi: Rawat Publications.

Chakravarti, Uma. 2000. 'State, Market, and Freedom of Expression', *Economic and Political Weekly*, 35(18): WS12–17.

Chanda, Ipsita. 2005. 'Kyunki Main bhi Kabhi Tulsi Thi: Opening and Using the Black Box of Primetime Telereality', *Journal of the Moving Image*, 4, available at http://www.jmionline.org/film_journal/jmi_04/article_04.php. (last accessed on 7 February 2012).

D'Cruz, P. and S. Bharat. 2001. 'Beyond Joint and Nuclear: The Indian Family Revisited', *Journal of Comparative Family Studies*, 32(2): 167–94.

Deshpande, Anirudh. 2009. *Class, Power & Consciousness in Indian Cinema & Television.* New Delhi: Primus Books.

Indiantelevision.com. 2003. 'Indiantelevision.com's Interview with Scriptwriter Shobhaa De', 11 February, available at http://www.indiantelevision.com/interviews/y2k3/writer/shobhade.htm (accessed on 10 February 2010).

Jaffrelot, Christophe. 1996. *The Hindu Nationalist Movement in India*. New York: Columbia University Press.

Jensen, Robert and Emily Oster. 2007. 'The Power of TV: Cable Television and Women's Status in India', NBER Working Paper No. 13305, Cambridge, August.

Krishna, Sonali. 2004. 'A Tale of 3 Ks', Indiantelevision.com, 5 November, available at http://www.indiantelevision.com/special/y2k4/3ks.htm (accessed on 15 March 2009).

Mankekar, Purnima. 1999. *Screening Culture, Viewing Politics: An Ethnography of Television, Womanhood, and Nation in Postcolonial India*. Durham and London: Duke University Press.

———. 2004. 'Television and Erotics in Late Twentieth-century India', *The Journal of Asian Studies*, 63(2): 403–31.

Marsh, Julie and Howard Brasted. 2007. 'Fire, the BJP, and Moral Society', in J. McGuire and I. Copland (eds), *Hindu Nationalism and Governance*, pp. 283–302. New Delhi: Oxford University Press.

McMillin, D. 2003. 'Marriages are Made on Television: Globalization and National Identity in India', in L. Parks and S. Kumar (eds), *Planet TV: A Global Television Reader*, pp. 341–59. New York and London: New York University Press.

Munshi, Shoma. 2010. *Prime Time Soap Operas on Indian Television*. New Delhi: Routledge.

Nigam, R.B.L. 2005. 'The Vicious Women of TV Serials', *Organiser*, 56(32): 41.

Niranjan, S., S. Nair, and T.K. Roy. 2005. 'A Socio-demographic Analysis of the Size and Structure of the Family in India', *Journal of Comparative Family Studies*, 36(4): 623–51.

Page, David and William Crawley. 2001. *Satellites over South Asia: Broadcasting, Culture, and the Public Interest*. New Delhi: Sage Publications.

Palackal, A. 2007. 'Review: Tulasi Patel, ed., *The Family in India: Structure and Practice*', *International Sociology*, 22(5): 575–8.

Puniyani, R. 2003. *Communal Politics: Facts vs. Myths*. New Delhi: Sage Publications.

Rajagopal, Arvind. 2001. *Politics after Television: Hindu Nationalism and the Reshaping of the Public in India*. Cambridge: Cambridge University Press.

Reddy, C.R. 2002. 'Religion and Fertility Behaviour: Canards and Facts', *The Hindu*, 10 November, available at http://bit.ly/15yxCsr (accessed on 10 January 2012).

Roy, Abhijit. 2008. 'Bringing up TV: Popular Culture and the Developmental Modern in India', *South Asian Popular Culture*, 6(1): 29–43.

Roy, Abhijit. 2011. 'Jassi Jaissi Koi Nahin and the Makeover of Indian Soaps', in P.P. Basu and I. Chanda (eds), *Locating Cultural Change*, pp. 19–53. New Delhi: Sage Publications.

Sarkar, Tanika. 1991. 'The Woman as Communal Subject: Rashtrasevika Samiti and Ram Janmabhoomi Movement', *Economic and Political Weekly*, 36(35): 2057–62.

———. 1999. 'Pragmatics of the Hindu Right: Politics of Women's Organisations', *Economic and Political Weekly*, 34(31): 2159–67.

Sarkar, Tanika and Urvashi Butalia. 1995. 'Introductory Remarks', in T. Sarkar and U. Butalia (eds), *Women and the Hindu Right: A Collection of Essays*. pp. 1–9. New Delhi: Kali for Women.

Savarkar, V. D. 2003 [1942]. *Hindutva: Who is a Hindu?* New Delhi: Hindi Sahitya Sadan.

Saxena, Poonam. 2009. 'Welcome to the Indian "Telly" War', *Hindustan Times*, 2 May, available at http://www.hindustantimes.com/Welcome-to-the-Indian-telly-war/Article1-406579.aspx. (last accessed on 7 February 2012).

Shariff, Abusaleh and Azra Razzack. (2006). 'Communal Relations and Social Integration', in A. Kundu, M. Dubey, N.J. Kurian, and R.I. Abbas (eds), *India: Social Development Report*, pp. 96–109. New Delhi: Oxford University Press.

Shivdasani, Menka. 2001. 'Stereotypes Galore', *Hindu Business Line*, 26 November, available at http://www.thehindubusinessline.com/businessline/2001/11/26/stories/102687b2.htm (accessed on 18 May 2009).

Subramanian, N. 2008. 'Legal Change and Gender Inequality: Changes in Muslim Family Law in India', *Law & Social Inquiry*, 33(3): 631–72.

Traister, Rebecca. 2011. *Big Girls Don't Cry: The Election that Changed Everything for American Women*. New York: Free Press.

Uberoi, P. 2000. 'A.M. Shah: The Family in India: Critical Essays', *Contributions to Indian Sociology*, 34(1): 129–31.

Unnikrishnan, Chaya. 2003. 'Vote for Your Favourite Vamp,' *Screen India*, 3 May, available at http://www.screenindia.com/old/fullstory.php?content_id=4463 (accessed on 3 February 2010).

Yechuri, Sitaram. 2009. 'Darkness at Noon', *Hindustan Times*, 20 October, available at http://www.hindustantimes.com/Darkness-at-noon/H1-Article1-466968.aspx (accessed on 21 October 2009).

11

Sange Thakun

Bangla News Channels and Media-citizenry

NILANJANA GUPTA

This chapter begins with the lament that while television in India has completed 50 years as a phenomenally successful medium, it seems that we—the academics and historians of contemporary media and culture—have been slow to endow the study of television the importance and the seriousness that it deserves. If we are truly and honestly to commemorate the success and almost bewildering growth of this medium, we need to construct pedagogical tools, analytical approaches, and institutional structures for recording (in both senses of the word) and understanding the practices that define the production and consumption of television in India. At present, while television is one of the most popular media in India, the study of television is, at best, regarded as a minor variation of film studies or, worse, seen entirely as an event which can only be considered as a cultural indicator of changing social values. Those who have written on television have largely depended on theoretical frameworks that may help to unpack the production and consumption practices of television in Europe or

North America, but often seem to ignore the particular issues and structures of television in India.

Narratives of the Regional

One feature that seems to mc to need a great deal of attention because it is a particularly Indian phenomenon—and that too a particularly fraught issue—is the fact that in India, many audiences will have the option of watching programmes in three different languages and that many members of the audience do, in fact, watch programming in these available languages. Most discussions on the audiovisual media in India seem to be concentrated on either the global or the national. Hindi cinema, as a site of national narrative for India, has provided an accepted framework for most analyses. While there have been some studies on the so-called 'regional' cinemas, this attitude still seems to be deeply ingrained, as argued in Sharmistha Gooptu's (2010: 12) book, *Bengali Cinema: An Other Nation*:

> In her book *Cinema at the End of Empire: A Politics of Transition in Britain and India*, Priya Jaikumar has made a case for abandoning the 'rubric of national cinemas' and abdicating 'the nation as an organizing device' in order to study the transnational politics of decolonization in British and Indian films made between 1927 and 1947. This is an important departure, for, as she notes, 'The framework of national cinemas has become a dominant analytical trope in Film Studies because of the nation's function as a central axis along which films are regulated, produced, consumed, and canonized.' Extending Jaikumar's paradigm *within* the Indian context, I argue that Bengal's cinema presents a history which brings to the fore the deeply-contested terrain of 'national' cinema, as also its ultimate subversion, and posits the creation of what I call the 'alternative imaginary of the Bengali film'. The existing scholarship on Indian cinema has been largely concerned with Hindi cinema made in Bombay, which is conceptualized as India's 'national' cinema.

In the case of television in India, the situation is even more lopsided as there is very little academic work which looks at the phenomena of regional language programming, though audience surveys show that regional language programming is extremely popular in the southern parts of India and other regions where the language primarily spoken by the population is not Hindi, such as the states of Maharashtra and Bengal. There has been a complicated and often bitter relationship between various languages and the dominant Hindi language in India. The issue has often been reflected in the political equations of the nation and has been particularly complex in the world of television. One reason for this is that before private television channels appeared, Doordarshan, the state-owned channel, was used as a means to popularize and—according to some critics—force Hindi into the states where Hindi was not the language of the people. All through the 1980s, all regional centres of Doordarshan, then the only television network in India, broadcast the so-called 'National Programme' in Hindi for fixed hours during prime time. This included the daily news broadcasts too. There was widespread opposition against what was seen as an imposition of the culture and politics of dominant groups of northern India. Exclusively regional-language programming was allowed in a second channel that was started in every region.

The experiences of the states where Hindi is not spoken are not all the same and indeed, even in Hindi-speaking states, the rise of private broadcasting has given rise to distinct regional programming. This is being reflected in the advertising patterns in recent years. As there are very few official sources available or academic works on such issues, one is forced to rely on websites meant for the industry. Most articles on such websites seem to be unanimous about the phenomenal rise in regional channels. According to one report, regional channel revenues across all genres grew at an annual rate of 26 per cent from 2006 to 2010, the year in which the revenue recorded was Rs 42 billion (Pandey 2012). Television in Andhra Pradesh was a case in point where advertising revenue was doubled from Rs 340 crores in 2006–7 to Rs 725 crores in 2010–11 (Reddy 2011). According to a report in April 2012, 55 per cent of the total advertisement volume on television was on regional channels, growing from 47 per cent in 2009 (Mitter 2012). Regional channels also command the highest share of overall televison viewership, with 33.4 per cent of viewers watching regional channels, followed by 27.4 per cent watching

Hindi general entertainment channels (Mitter 2012). All these figures and analyses emphatically argue that regional-language television is an extremely successful phenomenon all over India.

Regional News Channels

There are, therefore, many issues which need to be considered while mapping Indian television that hosts channels from across the world along with local-level programming. What happens when there is this multilayered structure of programme options and what is the relationship between these simultaneously existing options which are not necessarily mutually exclusive of each other? What kind of audience expectations are raised by them? Is there a hierarchical relationship implicit, based on their language?

A key factor in the growth of television in India has been the phenomenal increase in the number of 24-hour news channels. Table 11.1 charts the rate at which news television has grown in India.

At the end of 2011, the number of news channels in India was 402, that is, approximately 49 per cent of the total of 821 channels.[2]

Table 11.1   Consolidated List of News Channels in India, 2000–10[1]

Year	No. of news channels	Total no. of channels	% of total channels
2000	1	1	100
2001	39	44	89
2002	15	24	62
2003	12	24	50
2004	10	28	36
2005	10	15	67
2006	28	39	72
2007	39	74	53
2008	59	152	39
2009	33	79	42
2010	22	47	47
(2000–10)	268	527	51

Source: Roy (2011: 762).

Channeling Cultures

At the end of 2010, regional channels counted for 50 per cent of the total number of news channels in India and surely continue to be more in numbers than the 'national' (Hindi and English) group of news channels (Roy 2011: 766).

Using an ethnographic approach and focusing on the news channels' policy of engaging 'stringers', Srirupa Roy (2011: 774) argues that the '...expansion of the television news industry over the past two decades has had a "provincializing effect" of enabling various non-metropolitan elites to seek, and in several cases gain, class and status mobility'. Roy (2011) has also explored the implications of this development in the domain of political practice and has suggested that the growth of television news has accompanied, and even enabled, the proliferation of 'back-channel' or crony politics in local and regional political arenas. Somnath Batabyal (2010: 398), in his comparative study of the Hindi STAR News and the Bangla STAR Ananda, argues that 'Private news television in India...is a product of affluent journalists and their managers, produced for their well-to-do, privileged audience and institutionalising an imagined nation on the verge of global leadership'. Simon Cottle and Mugdha Rai (2008) use their concept of 'communicative frames' to compare the public broadcaster Doordarshan's news programming with that of the private channel, NDTV 24×7. They argue that the two offer different modes of presenting news, as the self-perception of their particular roles is different. But, in general, news channels in India aspire to intervene into the processes involving democratic institutions. I quote:

> Today's cacophony of conflicting interests in diverse civil societies, including India's, increasingly seek to mobilize their claims and aims in the media sphere. These same forces of change have given rise to a growing disenchantment with notions of democracy conceived in largely procedural and electoral terms, with calls for a 'deepening' and 'democratizing' of democracy within agonistic and pluralized civil societies... (Cottle and Rai 2008: 78)

This chapter situates itself within this changing scenario and attempts to theorize the shift from a discourse of deliberative democracy, largely controlled by the upper classes and upper castes in India, to performative democratic practices of the people. India is seeing a devolution of power into the hands of social groups who were previously excluded from the processes of democratic discourse and this chapter will, taking the case of West Bengal, attempt to argue that regional television news channels are both part of and partaking of these shifts in democratic practices.

Maharashtra, West Bengal, and the four states in the southern part of India have the largest number of 24-hour regional-language news channels. This means, of course, that the viewers in these regions have several kinds of news available to them at any point of time: those of the global channels; the national Hindi and English-language channels; and the regional-language ones. However, the various histories and political configurations of the states or regions bring in great variety in politics and ideologies of the television in different regions. For example, a discussion on regional-language news channels in the southern states on Indiantelevision.com website seems very disapproving of the current trends: '[T]here are currently 14 TV news channels in Andhra Pradesh. Except for two or three channels, all the rest are directly or indirectly controlled by politicians or their proxies. It's an open secret that most of them have officially become the tools of political agendas. One wonders why should they be called news channels at all?' (Mishra 2011).

The author discusses Tamil Nadu as a state where politicization through control by family business and political party has already reached a saturation point and anticipates that the two states of Karnataka and Kerala are on the same track too. There is a wide variety in patterns of ownership and control across different regions in India, making it difficult to generalize. The way audiences consume or use news is, in itself, an interesting issue as is the question what constitutes or is defined as news. For regional channels, there is the additional need to create a separate niche for themselves.

Transmitting Democracy

Each region has its own unique historical trajectory which is responsible for the development of a regional identity that is subsumed into the larger narrative of the nation. For example, the effort of a couple of Bangla channels to consciously aim at attracting the audience of the neighbouring country, Bangladesh, in addition to that of West Bengal, slightly complicates the features identified with an exclusive regional identity within a nation. However, this chapter will only look at some key features of 24-hour news channels in Bangla language in West Bengal alone. Within this limited field, the point of focus would be to identify the processes of construction of a kind of region-based political, cultural, and social citizenry which is distinct from the larger national construction of citizenship that is created by what is often called the national–popular.

There are, at present, 10 24-hour news channels in Bangla and many other general entertainment channels that telecast news and discussion on current affairs.[3] In West Bengal, the first programme to really create a model for regional news programming was a half-hour programme called *Khas Khobor* (Important or Significant News), which was created by a private producer and started on Doordarshan Calcutta (DD1 and DD-7) in 1998. This was a hugely popular programme that redefined the concept of news for regional Bangla-language viewers. Combative in nature and claiming to uncover local scandals and improprieties, it used a style which emphasized sensationalism over analysis and used the camera in outdoor locations in a new way. Reporters would visit locations and the literal 'the man on the street' would be interviewed. Usually, these location interviews were juxtaposed with formal interviews with police, government officials, or other 'authority figures'. Interestingly, the interviewees chosen from among the public would usually be much less articulate and their language would be less sophisticated, yet when juxtaposed with the formal statements of the officials, the level of emotion expressed by the 'man on the street' would serve to 'authenticate' the voice of the common man. The success of this segment perhaps helped to shape the 24-hour Bengali news channels which later came into being. One of the officials of the NE Bangla channel, in a private conversation, characterized his channel as the 'poor man's channel',[4] and it seems that this is true not only of NE Bangla, but of many other regional-language channels as well. The phrase itself is loaded and lends itself to several, perhaps simultaneous, readings. The person used it to refer to the fact that the channel did not have the funds and resources necessary to have a genuine newsgathering and news analysis bureau and that what it offered was 'poor' when compared to the resource-rich productions of the national networks like NDTV or the Times network. However, the phrase can also be read as suggesting that this channel caters to the 'poor man', which raises the issue of whether there is a distinct audience segmentation at work between the readily available international channels such as BBC or CNN, the national English news channels such as NDTV and CNN–IBN, the national Hindi-language news channels such as Aaj Tak or Zee News, and the proliferating Bangla news channels. While important national news or sport news does get coverage, the focus of these channels is almost exclusively on events in the state of West Bengal. It is perhaps relevant to mention the fact that though literacy rates are rising all over India, including West Bengal, and though there has been a rise in the number of regional-language newspapers too, television is still a primary source of news for a large segment of the population.[5]

One significant trajectory in media theory looks at journalism as a key player in the social reproduction of order and stability:

> ...journalists join with other agents of control as a kind of 'deviance-defining elite', using the news media to provide an ongoing articulation of the proper bounds to behavior in all organized spheres of life...[they play] a key role in constituting visions of order, stability and change, and in influencing the control practices that accord with their vision...In sum, journalists are central agents in the reproduction of order. (Ericson *et al.* 1987: 3)

Many analysts have pointed towards the means through which such ideological operations gain force: 'Journalism, then, pervades society with visualisations of order, coherence and unity, which are verified by reference to the eyewitness ideology of newsgathering, and by the ocular proof of visual evidence...' (Hartley 2008: 921). What these analyses are suggesting is that news is primarily devoted to representation of aberrations, disruptions, and exceptions. The government works honestly—that is the norm. News is created when there is a scam. Or, traffic flows smoothly; only accidents are 'news' and so on. This framework of analysis, that seems to be shared by most commentators on television news journalism in Britain and the United States (US), concludes that television news inevitably upholds the existing power structures and legitimizes the power elite in society. In doing so, news journalism actually disempowers the average citizen who becomes visible only when he/she is the victim of crime, or natural disaster, or some other 'disorder', while the screen is dominated by visuals and interviews of the 'deviance-defining elite', that is experts, officials, and analysts. This framework of analysis may hold true for the national-level English-language news channels in India, which generally follow this pattern of news construction; however, this chapter suggests that the Bangla-language news networks have evolved a different concept of news and representative mode.

In this age of the image, as opposed to the modernist world of the word, the news journalist becomes a distorted version of Walter Benjamin's *flaneur* figure. With camera and mike as tools, the television journalist, while creating a reality consisting of the everyday occurrences around him/her, emplots contemporary history. This emplotment has two aspects: narrative structure and a structure of feeling. The television journalist becomes the mediator through whose lens we interact with and comprehend everyday reality. This is more true for the news creators of the regional channels than the national ones who have the narrative of the nation as an overarching structural matrix. The local or the regional newscasters, who are working on the streets of the city or the pathways of the village, take the small events of the local—water logging in a particular area, hostel students protesting bad food, a child abandoned in a government hospital—and use the indignant voices of 'the people' to create a patchwork of re-presented reality.

In the case of the Bangla, and broadly regional news channels, an apparent disadvantage is turned into an advantage. Reporters are on the streets and pathways, and the interviewees are invited to speak in an apparently random manner. Instead of the larger issues of state and nation, what are made central are the 'little' stories of the 'little' people. The relatively simplistic format of the news stories too adds to the sense of immediacy and authenticity. In the national channels, a story is edited, file pictures are added, facts and figures are provided, and an issue is firmly placed within the larger framework of the nation-state. The 'poor man's channel' takes the story and represents it in a relatively 'raw' fashion, thus emphasizing the sense of the 'real'.

This narrative of the real, created by the regional newscasters, has an almost eerie resemblance to the existing modes of popular film in Bangla. In an earlier paper, I had attempted to identify patterns of media consumption by asking individuals about the range of media products they viewed or read, in order to see whether there was a co-relationship between different aspects of the consumption basket (Gupta 2005). In this survey, it was found that the segment of respondents who spoke only Bangla and watched almost exclusively Bangla news channels, also watched almost exclusively the popular Bangla film, both on television and in the cinema halls. That study conducted discussions on media consumption and tried to gauge audience reception by using pre-recorded segments from news stories in the discussions. The target groups were women whose literacy competence was very low or nil. The structural, narrative, and ideological similarities of the two forms are striking. Both use the form of melodrama where the voice of the 'ordinary', socially unprivileged individuals or groups are privileged and the voice of the rich, the powerful, and those with administrative power becomes the oracle of evil. What is perhaps even more interesting is that the visual language of the two forms also shares a great deal in terms of framing of the shots, camera angles, etc. For example, in a film called *Tomar Rakte Amar Shohag* (Ram Mukherjee, 1993), the hero, a worker, was protesting the exploitation faced by them and led a group to talk to the manager. The placing of the characters, the framing of the shot, and even the gesticulation was uncannily similar to a story on workers' protest shown on Alpha TV (later Zee TV) (Gupta 2005). Beyond the fundamental level of the melodramatic form, they also seem to share certain aspects in mode of representation and ideology.

In the moral economy of the typical melodrama, the existence of powerful emotion is a virtue in itself. In most of the Bangla newscasts too, the fact of the existence of a strong emotion is often shown as paramount in establishing the 'rightness' of a position, while the legal or more logical position is either left out of the narrative or is, it is implied, 'wrong'. A popular and often repeated storyline on the Bangla news channels concerns an abandoned child being brought up by a woman who then gets into trouble for not having any legal right to the child. Sometimes, this may be a nurse in a hospital who took home an abandoned child from her workplace to bring up as her daughter, or a family who found a child abandoned in a shop or on a street and took her home. When they try to admit the child to a school and cannot produce a birth certificate or any other legal document, the 'do-gooder' gets mired in legal problems, and often the child is taken away from the care-giver. In such stories, the archetypical emotions of the 'mother' overshadow every other point, including the fact that what she did was legally untenable, while the authorities are shown as the big, bad men who come to tear apart a family. There is no mention at all of the legal provisions, their admitted shortcomings, or even a mature attempt to familiarize the audience about the proper procedures. Instead, the emotional right of a mother is highlighted and established through close-ups of her anguished tears, and made part of the larger narrative modes of the channels' tendency to show the moral integrity of the common man and the unresponsive and often corrupt administrative structures that may be legally right but are clearly morally wrong.

The predominance of the narrative mode of melodrama is heightened by another characteristic typical to the form of television—the creation of the intimate. Not only can the camera use close-ups to highlight the expressions of emotions of the individuals, it can enter and capture for the audience the most intimate of spaces—the home. In the coverage of the murder of a young boy on 15 February 2011, who was escorting his sister home and was murdered while trying to protect her from the unwelcome attentions of a group of local thugs, the camera ruthlessly showed the grieving family, almost constantly, in their small cramped home with relatives and friends crowding the room and the mother in the centre, overcome with grief. The loud wails of the mother and the shocked expressions of the girl who was attacked became the focus of the story as the camera recorded these scenes interminably. Emotion itself and its manifestations seem to be the subject of the story overshadowing the facts and analysis of the case.

In the run-up to the 2011 assembly elections in West Bengal, both the major regional channels, 24 Ghanta and STAR Ananda, had several hours of programmes each day where the camera kept moving around, entering homes and workplaces, with celebrities or experts accompanying the journalists, as people were asked about their concerns and their opinions. Again, the intimate as a mode of representation of news programming set it apart from the more formal modes that are seen in the programme format of national-level channels.

Most of the Bangla channels attempt to create a mode of addressing the audience which is informal. One way this is done is in the way the on-spot reporter speaks to the anchor in the studio. The reporter addresses the anchor with the suffix *da* or *di* (brother or sister) and the story is presented almost like a conversation the reporter is having with the anchor which the audience is privy to. The anchor is usually more formally dressed in a suit and tie or, if female, in formal Indian or Western wear, while the reporter usually is seen in informal clothes. The anchor's language and mode of address to the audience is more formal and often, the anchor summarizes the story that the audience has just heard in a more formal way. In this layered presentation, there is the man on the street, the reporter, and the anchor, each of whom helps to mediate for the audience the immediate, intimate moment.

The institution of the 24×7 live news channels seeks to provide to its audience the illusion that there is a surveillance system in place in the competent, fearless hands of the news channel and so, the citizen can rest in peace because the channels will ensure that justice is delivered. This illusion is enhanced by the way viewers are invited to participate in the creation of news and views through strategies such as citizen journalism to short messaging services (SMS) polls to live phone-ins. It has become established, beyond question, that our duties as concerned citizens consist of collaborating with the news channels to create a 'proper', 'comprehensive' construction of the events of everyday life.

Media-citizenry

All these factors are, I would like to argue, creating a new politics where emotion, intimacy, the values of the melodrama, and the creation of audience complicity are coming together to define a new form of media-citizenry. While these features may be common for all levels of news channels, some of them are more accentuated in the case of regional-language news channels. For example, in the case of Bangla news channels, the melodramatic mode of representation creates its own logical structure in representing the news stories. In popular melodrama, the contradictions are stark and monochromatic, characterizations are one-dimensional, and the language that is used is always of a heightened pitch. The news stories that are aired on the Bangla news channels are also stark and one-dimensional. If two news segments on the same issue from a national English news channel and a regional news channel are compared, it will be seen that the national English news channel will use a more complex story structure with several subplots, interviews inserted, and often commentaries will be edited in, along with stock footage related to the issue. In contrast, the Bangla regional channel will show the same story in relatively stark and non-complex story structures. Often, the anchor will introduce the story in a way that practically tells the viewer how s/he should process the story.

I am not suggesting that one form of representation is necessarily better or worse. In fact, the voice of the ordinary person needs to be heard in any society that calls itself a democracy and it is important that the vernacular channels are the medium of such articulation. The framing and the structures within which these voices are inserted seems to be creating a distinct vernacular politics of citizenship which I have called media-citizenry earlier. The word *citizenry* rather than *citizenship* is being used to distinguish between a political identity which is determined by the political structures of the nation-state and that which is being created and encouraged by the media-driven inclusionary political framework. This new form of media-citizenry is not by any means to be dismissed as trivial or with no impact in the real world of politics.

It is, in fact, precisely this power of the vernacular that lends its charge to the local issues, such as that of land acquisition and development in Singur where Tata Motors was building a small car factory until local opposition grew into a mass movement and the factory was shifted to Gujarat. This local issue was catapulted onto national political scene and intense media coverage, particularly through the whole of 2008, played a big role in this process. Yet, even during the coverage of the Singur issue, the distinction between the concerns articulated by the vernacular channels—and by implication, those of the vernacular media-citizenry—and those articulated by the national English channels was clear. While the stories of the poor and the marginal dominated the Bangla television channels, the national channels focused more on the need for reforming the Land Acquisition Act and compensation packages.

Another example of the potential for mobilization and the evolution of a political media-citizenry can be found in the response to the cyclone Aila that hit the Sunderbans on 25 May 2009. The impact on the audience was great as was testified by relief agencies who commented that they had never seen such response from across the state. The coverage of the suffering of the local people by the regional channels had a very emotional impact across the state. Here, too, the coverage of the suffering undergone by 'ordinary people' was the key. The individual stories of loss and hardship were more overwhelming than the vast monetary figures provided by the print media about destruction and rehabilitation.

John Hartley (2008: 925) argues that journalism suffers from a 'twin commitment to truth and communication'. What he means by this is that journalism has, on the one hand, a commitment to hard facts and, on the other, a need to communicate by using fictional modes, file shots, editing, and other devices of what he terms 'fakery'. Hartley does not use this word with any degree of moral disapproval; he argues that the process by which communication of truth becomes possible is by choosing the most expressive visual to convey the message. One of the examples he uses is from a news story about a building burning down. In the news segment, a woman is shown on the street, clearly anxious and distraught, as she watches the rescue operation. The voice-over identifies her as one of the several people who are waiting for news of their family trapped in the building. Hartley says that the woman was, in fact, waiting for news about her pet cat. Yet, he claims, this 'fakery' does not negate the truth of the story. It is merely a question of choosing the best visual to communicate the message about the fire (Hartley 2008: 921).

Television news, with its reliance on the visual, is thus always attempting to capture the eye of the viewer through its pictures, and the most captivating is not necessarily the most 'factual'. However, the use of visuals must always at least seem to be authentic. In this process, the role of the news reporter is crucial as s/he provides the legitimizing link between the viewer and the experience being broadcasted. The authenticity of the experience is proved by the very presence of the reporter. The power of the reporter is, in turn, validated by our belief in the power of the eye-witness account. The camera provides the viewers opportunities to be eyewitnesses to the experiences being reported. Typically, the reporter will ask the camera person to show a particular shot, thus strengthening the idea that the viewer is as much an eyewitness as the reporter. The ability of the regional channels to cover the entire state almost instantaneously implies that the viewer too is able to know of any incident immediately and that too as an eyewitness.

These notions are strengthened by the fact that many of the channels almost constantly run a ribbon at the bottom of the screen with the names of the reporters, their areas of responsibility, and their mobile numbers. Some newscasters even directly address the viewers inviting them to note the numbers and call if any event occurs in their locality. All these point to a process of legitimization of the vernacular experience and the emergence of a different notion of political and ideological engagement on part of the individual, mediated by the role of the regional news channels.

* * *

In recent years in India, the political landscape has been dominated by regional parties or by parties who, while technically pan-Indian, are significant in limited areas only. The Bahujan Samaj Party and the Communist Party of India (Marxist) are both examples of the latter. It seems unlikely that even the national political parties such as Indian National Congress or the Bhartiya Janata Party will, in the near future, be able to form a government at the centre without support from regional parties. In these circumstances, the power of the parties is quite clear. But equally, if not more, significant is the fact that if regional television is an influential factor in determining and defining the ideological discourse of the region, then the quality of political discourse itself may be transformed to an emotion-based melodramatic discourse where the nature of the polity itself could be undergoing a fundamental change.

Notes

1. This table, sourced from the Ministry of Information and Broadcasting, Government of India, is cited by Srirupa Roy (2011: 762). These are channels that have been granted permission to uplink from India, and may not necessarily be 'active channels', that is, those that are actually on air.

2. Available at http://www.mib.nic.in/ShowContent.aspx?uid1=2&uid2=84&uid3=0&uid4=0&uid5=0&uid6=0&uid7=0.

3. List of permitted TV channels as on 6 March 2012, published by the Ministry of Information and Broadcasting, Government of India, available at http://mib.nic.in/ShowContent.aspx?uid1=2&uid2=84&uid3=0&uid4=0&uid5=0&uid6=0&uid7=0 (accessed on 23 May 2012).

4. Private conversation with an official of NE Bangla (who did not wish to be named) on 23 April 2009.

5. See Srabantika Basak (Neogi) (2011) for the detailed figures at the end of 2010, establishing television as having a higher reach than any other mass media in India.

References

Basak (Neogi), Srabantika. 2011. 'Indian Readership Survey—An analysis by RK Swamy Media Group', 23 March, available at http://www.exchange4media.com/news/story.aspx?Section_id=5&News_id=41511 (accessed on 8 December 2011).

Batabyal, Somnath. 2010. 'Constructing an Audience: News Television Practices in India', *Contemporary South Asia*, 18(4): 387–99.

Cottle, Simon and Mugdha Rai. 2008. 'Television News in India: Mediating Democracy and Difference', *International Communication Gazette*, 70(1): 76–96.

Ericson, Richard V., Patricia M. Baranek, and Janet B.L. Chan. 1987. *Visualizing Deviance: A Study of News Organization*. Milton Keynes: Open University Press.

Gooptu, Sharmistha. 2010. *Bengali Cinema: An Other Nation*. New Delhi: Roli Books.

Gupta, Nilanjana. 2005. '"Reading" Television: Towards a Definition of Audio-Visual Literacy', *Journal of the Moving Image*, 4, available at http://www.jmionline.org/film_journal/jmi_04/article_08.php# (accessed on 17 July 2011).

Hartley, John. 2008. 'Heliography: Journalism and the Visualization of Truth', in Michael Ryan (ed.), *Cultural Studies: An Anthology*, pp. 917–35. Malden, MA: Blackwell.

Mishra, Mahendra. 2011. 'Politicisation of TV News Content in South India', 2 February, available at http://www.indiantelevision.com/special/y2k11/Mahendra_Mishra_Yearender.php (accessed on 18 December 2011).

Mitter, Sohini. 2012. 'Regional Channels Gain as Broadcasters Chase Advertisers', 12 April, available at http://www.financialexpress.com/news/regional-channels-gain-as-broadcasters-chase-advertisers/935785/0 (accessed on 27 April 2012).

Pandey, Punit. 2012. 'Pure Music, Regional Key Trends', 18 January, available at http://www.indiantelevision.com/special/y2k12/punit_pandey_yearender.htm (accessed on 31 January 2012).

Reddy, Sanjay. 2011. 'Telegu Television Rewrites Script in 2010', 19 February, available at http://www.indiantelevision.com/special/y2k11/sanjay_reddy_Yearender.php (accessed on 31 January 2012).

Roy, Srirupa. 2011. 'Television News and Democratic Change in India', *Media, Culture & Society*, 33(5): 761–77.

12

Tears, Talk, and Play

A Window to Gender and Sexuality on Tamil Television

UMA VANGAL

The 'Sun'[1] rose on the satellite television horizon in the year 1993 and immediately captured the attention of an audience tired of Doordarshan and its centralized programming. The perception of Doordarshan as a platform for imposing Hindi-language content was firmly entrenched in Tamil Nadu, though the Tamil dubbed versions of some Hindi soaps such as *Junoon* were slowly making inroads into the local market. Sun TV's arrival was closely followed by channels like Jaya TV and then Vijay TV, and soon they captured the Tamil imagination across the globe. Today, Tamil television consists of over 80 channels, including the ones in Tamil Nadu, Singapore, Malaysia, Canada, Sri Lanka, and Internet television. This list includes Doordarshan channels and private channels owned by individuals, media conglomerates, and political parties. These channels offer a wide-ranging fare from entertainment to devotional, cartoons, and music, from news and film/film-based content to evangelism and political programmes.

Owing to the very nature of the televisual medium and the multiplicity of programmes, channels, and means of reaching out to audiences, it is difficult to study television programming unless this is done in a specific geographical and temporal context. And as such, Tamil Nadu is unique in that the audience is well versed in the audiovisual medium. As an audience, the Tamil people have already been brought up on a diet of cinema that became the site of public discourse. Television in the state carries with it the legacy of such a public domain. To this end, this chapter seeks to examine Indian television, more specifically Tamil-language television, over the last decade (2000–10), paying particular attention to representations of gender. I will analyse a cross-section of programming to ask if Tamil television has promoted figurations of strong women and expressed alternative sexualities. Using cultural studies and feminist studies approaches, and applying techniques of content analysis, textual analysis (both visual and discourse analysis), and reception studies, an attempt will be made to study gender issues and alternative sexualities in Tamil soaps and reality shows, and the public debates on them.

Tamil television channels' staple diet consists of films, films-based game shows, mega-serials, game shows, talk shows, talent shows, film-based countdowns, chat shows, and film reviews. News channels and music channels garner good viewership but the mainstay for advertisers in terms of revenue is the entertainment channels. The fact that soaps in Tamil Nadu are produced predominantly by women and cater to women (supposedly) makes it necessary to take this phenomenon seriously. They all feature strong-willed women protagonists who triumph over all obstacles placed in their paths by callous husbands, villainous family members, sundry business rivals, and generally, a host of people who seem hell-bent on making things tough for them. The men often emerge as almost cardboard figures, invisible except as supporting characters. Television drama is characterized by this focus on indomitable women, emphasizing their emotional and psychological growth through a clever placement of moments of dramatic intensity. As such, they draw heavily on the genre of women's melodrama. Reality shows, too, fall back on the same elements of dramatic intensity, generating strong emotions in the audience.

As elsewhere, Tamil television utilizes melodramatic narrative elements, gesture, expression, and movement in both soaps and reality shows to connect with audiences.

Most soaps on Tamil television feature women protagonists who triumph over all odds—they cry, scheme, and win by any and all means. Overtly the codification of women through images is repetitive, stereotyped, and regressive— these women are depicted as submissive and yielding to demands and threats, from dowry to domestic violence to suppression, without a murmur. Alternatively, they are portrayed as shrews who need to be tamed or the so-called modern woman who is aggressive, scheming, and quite often unfeminine. The moderate women are few and far between.

Women on Top

In Tamil television, there is no monopoly of a single production house. These production houses are allotted slots and produce a variety of television programmes—soaps, game shows, reality shows, and so on. In-house channel productions are usually the newscasts, talk shows, features, interviews, phone-ins, and devotional and music programmes. Interestingly, several of these production houses are dominated by women who are either owners or executive and creative heads entrusted with conceiving, writing, producing, and directing women-oriented soaps. Kutty Padmini, Radhika Sarathkumar, Khusbhoo Sundar, Ramya Krishnan, Sujatha Vijaykumar, Subhaa Venkat, and Kala Master are, just to name a few, prominent women television producers. Their work is not restricted to soaps but includes reality shows, talent shows, cookery shows, and game shows. In addition, there are film companies that have diversified into television production such as AVM, Vikatan from the Ananda Vikatan Media group, Gemini Productions, and Min Bimbangal. The two men who stand out among all these women, Thirumurugan and Thiruselvam, both graduates of the Film and Television Institute of Tamil Nadu, are the successful directors of two of the longest-running serials on Sun TV, *Metti Oli* and *Kolangal* respectively. These figures became household names in the state since they both directed and acted as the male counters to the female protagonists in soaps that are noted for a realistic, down-to-earth treatment of the issues plaguing women.

In reality, Tamil Nadu might have the largest number of women chief executive officers (CEOs), usually in family-run concerns, from Tractors and Farm Equipment Limited (TAFE)[2] to many industrial houses in Coimbatore, and role models such as Indra Nooyi of PepsiCo, but the middle class largely aspires for steady government jobs for their daughters. Of course, over the last decade, parents do accept the information technology (IT) sector as an option. In Tamil mega-serials (long-running soaps), the woman does not hesitate to divorce a man when he does not treat her well. Women choose careers as entrepreneurs in diverse fields, from travel agencies to real estate, from software companies, digital copying companies to international business. In interior Tamil Nadu, such career options for young women have now become acceptable. This creates positive role models for many young girls who are in rural and semi-urban areas.

Tamil television, as elsewhere in India, is a preferred option for many yesteryear film stars, especially those who find good roles hard to come by. What is unusual here though is that all these actresses have turned producers who run production houses. Therefore, most often, it is women who are conceiving these characters as strong, bold, courageous women. While they are depicted as strong-willed women, they always seem to need the prop of a man. Though sometimes they draw strength from mothers, sisters, and female friends, the catalyst for their success is usually a man. In *Selvi*, her friend Anwar helps her establish her travel agency; in *Kolangal*, Tholkappian helps Abhi achieve enormous success; in *Kasturi*, her father-in-law stage-manages her metamorphosis; and in *Kalasam*, the manager is instrumental in shaping her revenge plan. In *Arasi*, though she is far more self-reliant, her husband's solid support is her strength and the loyal team she works with helps in her professional success.

Domestic violence, dowry harassment, date rape, contract killings, acid attacks, murder and mayhem, divorce, child abuse, human trafficking, organ harvesting, international crime, terrorism are all tackled by women. Therefore, the content and the representation of gender roles are wide-ranging. It is rumoured,

Radhika once sacked the male director in her media production house for handling the subject of women's morality in an insensitive fashion.

The question relates to intertextuality and the manipulation of the strengths of the medium. The representations of characters on television followed the same pattern as Tamil films and theatre, with the women being traditional once they are married, with the characteristics a chaste Tamil woman is expected to have—fear, submission, shyness, and the deification of mother figures. Standard gestures and behaviour patterns are reinforced. Similarly, for the men and the villains, the codes and gestures are familiar and easily recognized by audiences. In dramatic turning points particularly, use of raised voices, intense intonations, and tearful expressions dominate. The melodrama format of the Tamil soaps presents a world where the women overcome the hurdles to leap from housebound submissive women to confident CEOs with ease. They do not seem to have everyday issues such as school, cooking, and catering to the needs of the family, except in a decidedly middle-class setting such as *Metti oli*. They seamlessly blend the home front and the business front. Parents and in-laws are supportive and even accept relationship of these women with their male benefactors. The problems they grapple with are a way of bringing in issues and agendas that the makers wish to inject into the narrative.

The production values of these soaps are fairly high as locations are strewn over India, Sri Lanka, and Malaysia. The camera work and editing is slick with a lot of gimmickry. Frequently used are low camera angles to give characters a larger-than-life image, especially in situations of conflict, and close-ups to register strong emotions and reactions when they are assaulted verbally, captured in evocative fashion to create maximum impact among sympathetic audiences. Exaggerated sound effects and white flashes are readily understood by an audience used to Tamil cinema. Multiple cameras, cranes, dollies, and tracks and round trolleys are used liberally to enhance the dramatic effects.

The costumes suit the mindset of the conservative yet modern generation of Tamil viewers. The older women are dressed in saris (silk or cotton as the occasion demands) with traditional *bindis* and jewellery (exception: Sudha Chandran in *Kalasam*). They eschew the revolting lipstick, bizarre bindis, and overly jewelled look of the K soaps.[3] The sets are as realistic as possible, though repetitive since most of them are shot in regular shooting locations. Make-up is the minimum required for the camera and in some cases, such as *Metti Oli*, absent to maintain the lower middle-class atmosphere and mood. But *Metti Oli*'s title sequence had a group of dancers *à la* Tamil films and this became very popular. Lighting is carefully planned and quite dramatic, especially in sequences with high drama and action.

And finally, to an audience that has lived on a staple of film songs, the title songs are as important as storylines. The songs are hummed by a huge number of people in the state and the popularity of the title songs is visible on YouTube. The title songs are selected with great care, lyrics written with a social context in them, and choreographed and filmed like music videos. The *Metti Oli* title track had group dancers, while *Selvi* and *Arasi* have the lead actor moving in a purposeful stride imbuing the characters with the inherent strength they seek. The *Kolangal* title song blends the dance, visuals of *rangolis*,[4] and of women and men walking and talking. *Kasturi* has the family on an outing and ends on the female protagonist. *Kalasam* shows us the mother and daughter turning to each other for strength and walking forcefully towards their destination. All the title songs are sung by popular playback singers and the producers invest heavily in outdoor locations. All these add up to credibility enjoyed by these soaps and heavy identification with the characters, the situations, and the problems faced by them.

As is common knowledge, the fan club culture is rampant in Tamil Nadu, especially among film audiences. The fact that most of the television protagonists are popular yesteryear film heroines ensures great curiosity around their appearance on television. These serials are so popular that spin-off merchandise are seeing heavy profits in the state.

Television had garnered dominant readings initially when the choice of channels was limited. Today though, the television audiences make meanings in a negotiated manner and at times, even in an oppositional manner. The intertextual elements from film, mainstream print, and other electronic media play a crucial role in the way these programmes are received.

The soaps follow such style and production techniques that Tamil audiences are already familiar with, due to their deep-rooted and long association with popular film and broadly popular culture. While *Metti Oli* appeals to the middle-class values and traditions held dear by Tamilians, and real problems faced by women in their homes, for *Kasturi*, it is the naiveté and innocence that works. In the case of *Selvi* and *Arasi*, untrustworthy and irresponsible men appeal among female audiences. *Kolangal* has been the longest of these soaps in terms of screen survival. Though, of late, audience have expressed resentment over this serial's unnecessary prolonging and bizarre tactics, they have nonetheless consistently enjoyed the situation of a man caught between two families, and the true test of loyalty that is won in the end.

In a state where the chief minister[5] openly flaunts his two marriages and many ministers follow suit, and bigamy is a well-kept open secret among affluent industrialists and entrepreneurs, most of the women in the soaps are party to bigamy. Quite happily, they are second wives, illegitimate daughters, the children of the abandoned mother, or at the very least, motherless and dependent on the father. In *Metti Oli*, the father is an honourable man who chooses to bring up five daughters without recourse to a second marriage. In fact, the director of *Kolangal* has explicitly stated that he wanted to depict the plight of a man with two families as a deterrent to men who may want to ape this trend. He also has shown Abhi's ex-husband as a murderer, killing father-in-law and wife, so that he can be a rich man and force Abhi to respect him and remarry him. In *Arasi*, the issue of illegitimacy is brought up, while in *Selvi*, the woman agrees to be the second wife due to extenuating circumstances. In *Kasturi*, the husband actually marries thrice and kills off one of the wives to make way for a new one.

The common thread through all these soaps is that these women are ill treated, abused, and quite often discarded by the men in their lives with active help from the mother-in-law, business partner, or some other member of the family. They begin to rely on this other male figure for advice and find it difficult to survive without their help. The respectability and societal acceptance will never truly be hers no matter what her achievements as a career woman are if she is single or divorced. Again, the necessity of a man in a woman's life is emphasized. The serials suggest that beyond all her ferocity when provoked, a true Tamil woman will be traditional, long suffering, strong willed and patient, and try to keep her marital status intact.

Talk About Me!

Tamil television has today diversified into many formats—from clones of Oprah to moderated debates, from staged fights to collective soul searching. *Kadhaiyalla Nijam* pioneered talk shows in the south, with people opening up the floodgates of abuses on camera. Couples fought, parents united with long-lost children, a man who had imprisoned his wife got his comeuppance—the list is endless. The title promised that these were real stories. What the show did was make people comfortable with the idea of airing private laundry in front of an audience. It opened up the possibility of public discourse on issues that were hitherto considered to be taboos. Then came a flurry of such shows, many of which flopped. The most talked about shows today are *Visuvin Makkal Arangam*, *Arattai Arangam*, *Neeya Naana*, *Ini Oru Vithi Seivom*, and *Andha Kaalam Indha Kaalam*.

Talk shows hosted by Visu took on new meaning and relevance with rural and small-town Tamil Nadu getting its say on state-wide television. Some of the most powerful speakers on his shows have been young girls who made passionate and bold speeches on varied topics such as alcoholic fathers, education system, career options, single-language policy, and safe sex. T. Rajender's *Arattai Arangam* is a clone of Visu's programme and the show has received a good response. *Neeya Naana* is a debate organized in studio between two groups, quite often men and women, young girls and young guys, children and parents, teachers and students, dealing with issues such as studies, morality, double standards, discipline, and love. *Ini Oru Vithi Seivom*, hosted by Revathi Shankaran, brings together women and young girls to discuss diverse topics. In fact, the slot is marketed as women's time ('our time') across the state. Women open up and discuss acquired immunodeficiency syndrome (AIDS), abilities/disabilities, sex education, restrictions on women's time and movement outside home, career, early marriage, lesbianism, oedipal complex, and so on. The fact that they are amidst their own kind helps to bring up issues which might otherwise be untouched. In *Andha Kaalam Indha Kaalam*, hosted by the popular comedian Leoni, people play their favourite 'in the good ol' days' game, comparing the best of the past to the degeneration of values these days. This programme deals with a wide array of subjects that appeal to Tamil audiences: old film songs versus new film songs, poetry and literature, acting styles, music, folk forms, new media, communication trends, fashion, behaviour, and so on. *Ippadikku Rose* on Vijay TV is a path-breaking programme in Tamil Nadu. Having Rose Venkatesan[6] as its anchorperson, this show has paved the way for public debates on alternative sexualities on television. In Tamil Nadu, where generally the men pride themselves on their masculinity and virility, and women are revered for their coy, submissive, and gentle demeanour, this programme's efforts to break the apathy and indifference towards the issues and struggles of the third gender are exemplary. The first episode presented a panel discussion among transgenders on their lifestyles and problems.

In terms of visual treatment, talk shows are mostly similar to soaps. The specialty here is that the close-ups, quirky camera angles, reaction shots, coupled with emphasis on visually capturing warm smiles, heated exchanges, and easy camaraderie shared by participants and hosts, are all tied up to the alliterative and rhyming nature of the Tamil language. The talk shows also exhibit a major fascination with flamboyant and bombastic expressions, echoing the linguistic jingoism that characterizes the state. Talk shows have changed the nature of public discourse in Tamil Nadu in a subtle way as they open up taboo subjects for debate, give an opportunity to air views frankly, and allow multiple voices and views to be heard.

Games People Play

Tamil game shows, with their constant innovations and fabulous gifts on offer, have become highly popular over the years. *Jackpot* on

Jaya TV, where Khusboo's designer blouses and her not-too-perfect Tamil presumably add to the unique selling point (USP) as much as the concept of the programme, has attracted women participants from different cities of Tamil Nadu. The show, derived from *Family Fortunes*, is today the most-watched game show in Tamil Nadu with a long waiting list of people wanting to participate. *Thanga Vettai*, a popular game show produced by Radaan and hosted by actress Ramya Krishnan, offered prizes in gold. Ramya's clothes and jewellery on the show were widely discussed in the public domain. Here, the designer and the showroom received special mention with details on the design of the sari and the jewellery. Interestingly, only families could participate and they were united in their efforts to find the yellow metal for 'investment purposes', as they would all claim.

Jodi Porutham revolved around verifying the compatibility of a couple and also tested them on how well they knew each other. Many issues were brought out, such as how much the husband would be willing to stand up for wife in the situation of family dispute, does the woman tend to create rifts in the family, do they help each other in the household chores, and how is their sexual life. In a society that does not believe in talking about the intimate details of married life or indulge in public display of affection, this programme was path-breaking in many senses. *Thirandhidu Sesame*, set against a glamourous backdrop with television actor Vijay Adhiraj as host, became very popular. The title song, which was huge hit with audiences, was sung by popular Tamil rapper and playback singer, BLAZE, and the host. The participants would unwittingly expose their innermost fears, ambitions, and attitude to life while debating their options, with the anchors constantly questioning/challenging them on their choices.

Maanada Mayilada is a dance talent show with teams vying for the top slot every weekend. Produced and hosted by Kala Master for Kalaignar TV, this show is a huge success entering into its fifth season. The celebrity judges are all yesteryear actresses who became popular for their nimble feet on the big screen. With this show, the audience began accepting intimate dance choreography and couple chemistry on television and sometimes, the couple would go to extremes to ensure victory. Themes explored include romance, infidelity, and various issues in gender and sexuality.

Playfully, these shows managed to give space for gender and alternate sexualities to be discussed in an open and non-aggressive manner and opened the doors for participants, especially women, to express their choices. They also let women explore emerging career options and showcase their knowledge, exhibiting the edge that the new Tamil woman has in the fields of business, education, and the IT sector.

The talent shows and reality shows are catching on in a big way on Tamil television. While it is a fact that they are helping identify the next generation of entertainers, talent shows for children are affecting the child adversely in many ways. Parents impose a big burden on their children to perform on such shows; channels thrive on the competition, energy, and drama; and judges seem to not know any child psychology and ridicule and speak harshly to the children.

* * *

As a whole, the present-day Tamil television looks interestingly poised in the tussle between tradition and modernity. Soaps have created positive role models for aspiring young girls by offering a wide array of career options hitherto seen as male bastions. But at the same time, these soaps rarely portray women as self-reliant (they seem to need a man to help them get ahead), reinforcing patriarchy and the prevailing political climate. Talk shows provide a platform to bring many taboo subjects into public discourse and have become a forum where multiple voices can be heard. Game shows also give scope to the marginalized groups to showcase talents, and provide an outlet for creativity, intelligence, and sportsmanship.

It is not to say that Tamil television has been solely instrumental in effectively generating new debates on gender and sexuality in the public sphere. The print medium with stories that cover these issues, the IT boom, the changing face of Chennai as an investment and industrial hub, the influx of non-resident Indians returning home, the slow conversion of conservative Tamil Nadu into a more modern state with gender-sensitization programmes and awareness campaigns, the aggressive anti-AIDS campaign that began about 15 years ago in the state—all these factors have definitely contributed to the phenomenon. But Tamil television is definitely to be counted as one of the significant change agents, in particularly opening a window of opportunity for gender and sexuality to take main stage in the public discourse in the state.

Notes

1. Sun TV is a Tamil-language Indian cable television station. It is the flagship channel of the Chennai-based Rs 16,000 crore network promoted by Kalanithi Maran. It was the first fully privately owned television channel in India when it emerged in 1993.

2. TAFE is a US$ 750 million tractor major, incorporated in 1960, at Chennai in India, in collaboration with Massey Ferguson, and is among the top five tractor manufacturers in the world. Mallika Srinivasan, director of the Amalgamations Group TAFE, is one of the most successful women CEOs in India.

3. 'K soaps' is the popular name for the soaps produced by Ekta Kapoor of Balaji Telefilms in Hindi. Almost all these soaps feature women in formal silk saris, loads of make-up, wearing the traditional 'sindoor' prominently, living in lavish mansions, and spouting dialogue about Indian culture and heritage. Usually, the women who play the negative roles sport dark lipstick, eye make-up, and bizarre-looking bindis on their foreheads.

4. Rangoli is a kind of decorative design made on the floors of living rooms and courtyards during Hindu festivals. For details, see http://en.wikipedia.org/wiki/Rangoli (accessed on 19 August 2012).

5. M. Karunanidhi, elected five times as Chief Minister of Tamil Nadu between 1969 and 2011, has three official wives (the first is deceased) and regularly brings in public, his wives and offspring. In fact, his political and literary legacies are constantly fought over by his children from his multiple marriages. This practice of many wives is emulated by many politicians and celebrities in the state.

6. Rose Venkatesan is the first transgender to host a show on Indian television, though the lesbian, gay, bisexual, and transgender (LGBT) community is not too impressed at her attempts to promote herself and commercialize the issue.

Afterword

On the Unexpected Parochialism
of Media Studies

ARVIND RAJAGOPAL

Thomas Kuhn coined the phrase 'normal science' to refer to the kind of 'puzzle-solving' work undertaken when scientists agree about the rules of knowledge formation to advance science within a given paradigm. When their paradigm is in crisis, however, only a scientific revolution can eliminate the cognitive dissonance between new observations and old theories, Kuhn wrote.

No revolution is imminent, in Indian television studies, however, although it is one of many academic fields in crisis at this time, according to this definition. If Indian television studies is a project in search of a common object, as is suggested by many of these papers, a history of the object might illuminate the stakes of this project. At least four of the chapters specifically focus on how studies of television are to be defined and understood (Asthana, Kumar, Mankekar, and Roy), but no unanimity is discernible in defining their subject. They are aware that, in the case of India, while certain historical events are agreed upon, and certain theoretical frameworks used to be agreed upon, there is little consensus beyond remarking that nothing is like it was, or was expected to be. That is a good point at which to begin.

We could say, more broadly, that the history of television is nothing like it was expected to be. In 1961, Raymond Williams could still believe the technology was part of the 'long revolution' that would make equals of the poor and the rich.[1] Precisely for this reason, governments feared its power and often monopolized its control. Several decades later, television has come to symbolize non-confrontational communication or even pacification, and that too without overt state supervision. The medium is domestic, like the familial contexts it occupies. Criticism exists on television, to be sure, but 24-hour programming, mass audiences, the profusion of channels, and the predominance of entertainment genres, taken together, render criticism either as comedy or as political marginalia. Viewers may imagine ethical utopias, but what television offers is, mostly, consumer choice.

In the meantime, a newer set of technologies have become sites of more pressing concern: the so-called 'social media'. But what is television if not a social medium? Television's one-way communication is presumably less social than interactive media such as cell phones, Facebook, and Twitter. However we respond to this mode of analysis, it appears that the gravitational pull of new media dissuades many scholars from posing questions about it, perhaps because new media are seen to answer both old and new questions about what constitutes 'the social'.

An influential argument made by Marshall McLuhan (2011) suggests a reason why. Old media become the content of new media; the form of the dominant new communication technology defines media environment in a given period, McLuhan has suggested. A medium like television can no longer repay analytical scrutiny to the degree it used to, if we accept this reasoning; the world has moved on. Digital and mobile media have taken television's place as the most visible sites of academic labour although there is, embedded within these new media, a possible accounting of television even within the terms of someone like McLuhan.

A brief discussion of broader trends in media studies, of which McLuhan is one representative, will be helpful in this context. There is obviously more at work than academic fashion here. At stake is not simply the object of television, but received accounts of western history, which are simultaneously universal and particular, often consigning non-western developments to relative obscurity. Drawing on recent history, one way to produce a genealogy for the missed encounter between the west and its others is through the Cold War period, when the idea of communication became a powerful inflection of this empirico-transcendental doublet, incarnated in 'media'. In this period, Marshall McLuhan was perhaps the most celebrated theorist to link media explicitly with a conception of globalization that could cut across political differences, while tacitly reassuring the west that it might yet prevail.

As is well known, McLuhan argued that with oral–tactile forms of communication created by broadcasting, the hitherto print-dominated West would be equipped to better engage with the world's non-print literate majority.

In doing so, he provided an ingenious argument about the power of media, linking concerns about social order and the fear of the crowd, and latent apprehensions about the power of communist ideology, with the possibility of utopian transcendence. If the medium was the message, and the message was something about neo-tribal togetherness, it implied that communist propaganda would be neutralized. Modern media might, in fact, serve as silent allies of the West in the Cold War. Communications would create a global village, McLuhan prophesied, betraying his North American naiveté about the village as a place of consensus and harmony. He thus folded media globalization into an account of the West as an eternal harbinger of modernity, forever transcending its shortcomings.

One reason that these assumptions were not explored is that McLuhan appeared to be above the fray at a time when arguments about the United States (US) intervention and superpower conflict were partisan. To criticize the idea of a global village was perhaps not the most urgent task of criticism at a time of counter-insurgency warfare, in Vietnam and elsewhere. At the same time, McLuhan offered cultural analyses that were easily repeated but less often understood. They registered the sense that something new was happening, even if people did not know quite what it was.

His motto, 'the medium is the message', in fact resonates with arguments by post-metaphysical thinkers who aim to eliminate the transcendental position granted to the subject in the Western philosophical tradition. The eclipse of the speaking subject by language itself (Heidegger), the death of the author (Barthes) or the decentring of the subject (Derrida), the mapping of the real, symbolic, and imaginary onto gramophone, film, and typewriter (Kittler), and the rescripting of human agency as a technological event, as in *The Gulf War Did Not Take Place* (Baudrillard), are variants on the persistent attempt to dethrone the sovereign subject, and demonstrate how forces beyond its ken grant the subject the illusion of autonomy.[2] Communication in this understanding, far from providing the means to an ideal speech community and human emancipation, can only confirm the impossibility of such aims and ideals.

Post-metaphysical philosophers' task of decentring the subject, intended as a rational critique of irrationality, could not in fact eliminate the subject. The subject of reason, as Descartes' Cogito underlined, was not reason itself; in fact, its irreducibility to reason guaranteed the incompleteness of subjection and the possibility of freedom. The result of this critique of the subject was to normalize the Western European history of subject formation, and render it into a set of assumptions that seldom needed to be specified, since, most philosophers assumed, they corresponded to the unfolding of the history of reason. We can recall in this connection, Alain Touraine's observation that rationalization and subjectivation are the two faces of modernity (1995: 207). Rather than bypass the question of the subject, modes of subject formation need to be historically situated. Relevant here is Dipesh Chakrabarty's (2000) influential call to provincialize Europe. Chakrabarty seeks to align the subject of any theoretical inquiry with its geographical and historical location, rather than identify one area with a transcendental subject.

One example will illustrate the kinds of problems that follow as a result of normalizing European experience. In a widely cited essay, Gilles Deleuze(1992) states that his account of 'control societies', where post-sovereign power produces multiple subjectivities regulated by protocols of flow and interdiction, has no connection to the circumstances of 'three-fourths of humanity'.[3] Meanwhile academic accounts of non-Western countries are replete with human subjects whose inability either to conform to Western norms or to be theoretically decentred in scholarship only lead to affirming their provinciality vis-à-vis the West against the best intentions of scholars.

As in history, so in media studies, scholars of regions outside the West tend to look for conformity with norms claimed for European experience; the rest is, to use a communicational metaphor, noise. In fact, I suggest, if all history is European, as Dipesh Chakrabarty has written, then media studies are American.

The idea of America as the medium of the media underpins the key concerns, not only of media studies, but more broadly of. In the beginning, all the world was America, John Locke wrote, in his *Second Treatise of Government* (Laslett 1970: 319). If the world had deviated from the American condition, Locke clearly saw it as his task to underline this as a problem. The ideas of natural law on which his argument about inherent rights of property in personhood are based, were adduced to a laissez faire theory of economic development, in which the modernizing state was a caretaker at best. In this understanding, tools such as communication media are a potential to be actualized, rather than merely an apparatus of information transfer or state rule. A brief historical discussion will clarify my claim here.[4]

The unique circumstances of US history led to extraordinary policies for the eighteenth and the nineteenth centuries, with state subsidies for the postal system, libraries, schools, and newspapers, and more unusually yet, an absence of censorship. These were undertaken in the attempt to strengthen national bonds within a nascent, far-flung republic. The US state expected that interaction between citizens could only strengthen its social base, which was not a typical assumption for the time. It reflected the reality of the government's reliance on settlers to extend the frontiers, battle the natives, cultivate the land, and build the nation in the process. Lacking a standing army and without much by way of revenue, the state chose to undertake an extensive mobilization of the citizenry. Subsidized education and communication were indispensable to this task.[5]

Now, by what means is social order to be preserved when communication is relatively free, and mediated through print and post? Such media eliminate the physical presence of the speaking subject, and render messages beyond the spaces where their reception can be regulated. One of the most remarkable observations Tocqueville makes in *Democracy in America* (two volumes published in 1835 and 1840; see Tocqueville 2011) is that the distinction between state and civil society in the US was fungible and highly contingent; the political culture of the state was on a continuum with popular culture. Equality was mandated by law, discernible in society and reiterated in everyday life. Governmentalization of state power was not only political–economic, but simultaneously communicational, and devolved onto white settlers and their progeny by means of the state bearing the cost of education, information, and early forms of virtual interaction via the post.

In the US, media emerge as a ventriloquism of state power in and through the marketplace. A parliamentarism of marketplace communications arises to complement it, with conventions about neutral and bipartisan newspaper reportage taking shape alongside partisan pamphleteering. The growth of media was promoted in the historical and political context of a republic that sought to ally society with a fledgling state, including giving citizens the right to bear arms, we should recall. The market was not opposed to the state but in concert with it, which is why, in the US, advertising matured more quickly than anywhere else. Advertising, although an interested system of communication, is granted its own citational authority and claims to speak to the common interest as if it were the state. Who is speaking, where and when, become relatively unimportant, while the message itself acquires the status of sanctioned communication and is endowed with iterative force.

However, the unique characteristics of the US state's influence on its media, including its racial, religious, and nationalist entailments, have been normalized through a familiar method of distinguishing between those data that conform to a given theory and those that are aberrant. When theory collides with reality, Marx once observed, theory wins. It is reality that is declared deficient instead (Marx 1844).[6]

Afterword

The theoretical subject of the media in media studies is American. He or she exercises communicative privileges as a natural right in the marketplace, and encounters the state as a discrete actor rather than as an historically contingent support authorizing communication. Advances in media technology appear as means to enhance human capacity, and media form is understood as containing the secret of this possible enhancement. Here lies the paradox in the quest to eliminate the subject: media can only mediate by means of subjectivation, that is, by addressing readers and viewers who are addressable.

Even if we set aside the philosophical problem of independent, self-adequate objects (how can we know they exist, absent a subject able to know?), 'media' arise only from conventions that designate certain objects as such. Furthermore, if 'the medium is the message', this depends on a subject who registers the message and activates its codes. The media are not, therefore, independent objects, but are co-constituted through the subject via whom media achieve their mediating effects. And as we have seen, the normative subject of media studies operates in a New World context where the individual is sovereign and the media are a support to that sovereignty. Such understandings tend to be encoded in perceptions of the media object itself, and appear to emanate as properties of a given medium itself. It is somewhat analogous to the way a commodity is a hieroglyph of the social relations that produce it; the commodity's fetish character confirmed in its lack of intelligibility. Distinct from the commodity fetish, however, the media object presents itself as incipiently intelligible, albeit via its user's intervention. It is therefore by presuming the media object's ready intelligibility that its fetish character, is reproduced. Hence arises the curious status of media knowledge as both self-evident and mysterious.

It is in the inter-war and post-World War II period that a different theoretical model arose most forcefully to challenge such an idealized view of the inter-mediation of state and society in the US. The Frankfurt School's critique of mass society, informed by the experience of European fascism, marks this historical moment, one that is, in fact, on a line of continuity with the early American solution to the problem of order. Every American citizen was both medium and message of governmental power. In the early republic, this power was experienced as cultural ethos, not as state directive (Tocqueville 2011 [1835, 1840]). In mass society, citizens acquired an intensive conformism resulting from a mode of power where the distinction between state and market was hardly relevant. But in the context of Cold War rhetoric, the default framework was one in which media growth and the expansion of freedom were aligned most clearly with the idea of America, even if not with the historical experience of the US. Technologically advanced media, according to this logic, held the possibility of furthering the project of enlightenment, even if they were enmeshed in systems of social regulation and political control.

If Marx thought theory trumps reality, we could say in this case, it is myth rather than theory: the myth of America as the idea of natural law and Rousseauian harmony, and as the ideal future and destiny of the world. Geography however is not incidental to this myth, but constitutive of it. The New World, existing far from the decaying structures of European feudalism, offered a 'blank slate' for the utopian projects of enlightenment philosophies, and continues to function as the transparent and non-material support for mediating enlightenment in imagination. Colonial genocide and slavery exist at the margins of this myth and do not fundamentally alter it, because for all of its flaws, this idea of America provides a guiding light for collective action.[7]

If the theoretical subject of media studies is American, then, such assumptions do not inevitably retreat into authenticity or indigenism. Rather, the issue is of temporalities built into theory but is off limits to historicization, especially for media critics. The challenge of understanding 'neoliberal' politics and culture, however, presents important problems that can most usefully be illuminated by referring back to early American history. 'Neoliberal', in this context, is a placeholder rather than a precise descriptor. It refers to the ascendancy of market exchange as a model of governance, alongside both more punitive and less redistributive state policies. The media are, for the most part in this period, non-state entities, but clearly enjoy state-like authority via the market. Unlike the classical period of American liberalism, however, the market is dominated by large corporations, which benefit disproportionately from state subsidies. The advent of neoliberalism then indicates the combination of appeals to the ideals of classic liberalism, from the state and the market both, with the prefix 'neo' to what remains to be determined, namely, the character of the links between politics and culture given the subversion of classic liberalism's claims of formal equity. No scholar can disagree that the work of the media is critical to the performance of neoliberalism. Indian media studies presents interesting questions because of the relative success of both commercial media and electoral democracy in India. Both the retreat of the state and market ascendancy obtain as well. What it crucially includes is a colonial history that ensures the importance of the state cannot be overlooked, as many of the chapters here have pointed out.

Indian Media: Locating an Emergent Field

Those who study television in India labour under a double burden of provinciality, in the view of many of those who study new media. First, they focus on the non-West, distant from where technologies incubate and advance, and second, they examine old rather than new media.

There is a familiar irony here. The political and theoretical vanguard in a field feel obliged to turn their backs on the majority (in this case, as in so many, the majority are also poorer) because their attention is fastened on defining tendencies and maturing contradictions premised on the centrality of their own situation as they understand it.

We can say that there are, implicitly, two kinds of media studies operating here, and although they are not clearly distinguished from each other, they reflect a salient divide. One follows the development of science and technology that leads to the development of media artefacts from paper and print to the telegraph, radio, and beyond. The historical location of this body of work is Western Europe and the US, because it is here that the work of the media can be assessed in terms of measurable deviations from the norm. Broadly speaking, the modes of subjectivation shaped by media, which would include cultural, linguistic, and religious factors, can be bracketed because they are normalized, as is the mode of state intervention. In these contexts, the medium can be granted agency ('the medium is the message') to the extent that the subjective mediation of the media can be treated as relatively homogeneous and taken for granted.

Meanwhile, studies of the media from elsewhere invariably have to negotiate multi-ethnic, multi-linguistic, and multi-religious contexts, where the state's presence is usually obtrusive, and seems designed to foil rather than foster the possibility of free communication. Accounts of the production, interpretation, or circulation of media texts, or of media organizations in their historical context, can consequently appear too detailed and inaccessible to all but the area specialist, even when such accounts are working through social scientific or critical theories. This is to some extent true across area studies, but the parochialism in media studies is more noticeable because, in fact, communications media themselves present a ready basis of comparison, as potentially modernizing technologies. However, media tend to scramble temporalities and present heterogeneous, non-linear outcomes, complicating any claims about modernizing effects. A frequent response is to use an anticolonial or national frame to subtend analysis, for example, with a national press, cinema or television, thereby bracketing historical specificity. It is striking how little distance has been traversed in comparative analysis towards, for example, asking how different state regimes are disposed towards communication technologies, or how different multi-lingual or multi-religious contexts accommodate and constrain media institutions and texts. The success of the best-received effort in this regard is, of course, Benedict Anderson's *Imagined Communities* (1983), and its success perfectly illustrates my claims here. Nationalism is presented as an empty modular form, to be filled through the combination of print capitalism and national sovereignty. Historical subjects, with their ethnic, religious, linguistic, and other contingencies, to say nothing of their concrete forms of thought, are incidental to his analysis.

Small surprise then that Indian television studies figures in specific disciplinary sites, including literary studies and anthropology, which emphasize ideographic forms of analysis, and that in these fields as well as in media studies itself, Indian television studies is a sub-field into which outsiders seldom venture and across which comparative arguments are seldom made. Perhaps the greatest interest however occurs in professional schools via fields such as market research, where input about reception and audience consumption is needed.

The result of this mode of partitioning inquiry is that it leaves intact the hierarchies posited between the West and the presumptively less modern non-West, when for more than a century-and-a-half, ideas, markets, and technologies have been tying them together ever more closely. For example, communications technologies appeared in the colony, as we know, principally as means of extraction, censorship, and surveillance, and thus of an authoritarian case of the relationship between media and state authority.

In this connection, it is relevant to note that the spread of technology was always claimed to incrementally, if not rapidly, Europeanize the provinces as sites of modernization. Forensic medicine, fingerprinting, and photography tracked and surveilled natives, while railways moved raw materials extracted from the hinterland to the port cities, and transported finished goods from the metropole. Communications technologies such as print, radio, and telegraph were rigorously controlled and subject to censorship, and the government maintained a monopoly over the airwaves to ensure their proper use.

Under the British, technology brought the provinces closer to Europe. It also held them apart as distinct entities, maintaining a rule of colonial difference. With political independence, the project of national developmentalism that followed the colonial state reproduced a raison d'etat aloof from popular sentiment as a condition of its existence. Rulers were now elected, but politics appeared external to the logic of technology and development both, and as a process more likely to subvert the rationality of governance than to advance it. It should be noted that, for most of its existence, television in India (which was inaugurated in 1959 and was opened to non-state broadcasters only in 1995) was insulated from competitive politics due to state control. However, many critics held state television to exemplify not the rationality of governance but its obverse. Why the government would invest in a technology that it proceeded to utilize such that even its propaganda was ineffective, was not clear. Indian media studies confronts precisely such inversions, where anticipations based on the experience from West and East bloc countries are confounded. Apparently, the result has been to dissuade from further inquiry by most scholars except those committed to a study of the area, however. It is for this reason that the main effort of my chapter is to pursue the questions raised by the authors in this volume about deprovincializing Indian media studies, and to critique the scholarly assumptions that reproduce such parochialism.

The Object of Televisual Criticism, and its Accounting

Television in India was introduced under the distinctive conditions of postcolonial development. High hopes were placed in the medium as a means of development, but great danger attended it too. Affluent middle classes thought of their own comfort, and had little patience for educational programming (Masani 1976). Businesses, for their part, saw little profit in adopting the state's priorities that held information and education above entertainment in programming. Meanwhile, foreign ideas and sentiments circled the subcontinent, looking for a chink in the armour of the nation's defences. Only the state could be depended upon to harness television for the collective benefit.

This was, of course, a view most often advanced by defenders of the state. Since the levels of investment required for televisual infrastructure exceeded what any business could afford at the time, and the domestic market was still limited, the idea of television as unavoidably a state enterprise in India seemed like common sense. The P.C. Joshi Committee on television in 1985 thus argued for 'an Indian personality' for Doordarshan, the state television system, assuming that while broadcasting might be independent, it had to be 'national'. How could it achieve that without state sponsorship, even if it sought to be independent?

This is all history today. Many would probably view it in Henry Ford's sense of the term ('history is bunk'). Samir Jain, vice-chairman of *The Times of India*, currently the largest-selling English newspaper in the world, is said to have remarked that if he were prime minister, he would ban the study of history, because he thought that history didn't exist.[8] When his brother, Vineet Jain, relaunched the Times' satellite news channel in 2006, it was called Times Now, conveying 'a breathless now-ness and immediacy, not leisurely features and analysis', according to 'The Times of Media', the company's official history. 'It is about creating the illusion of breaking news, even if it is in fact news that's already been broken' (Auletta 2012: 58).

Having decided that history doesn't exist, the Bennett, Coleman Company Limited, India's largest media conglomerate, is in a position to act on it. Newscasters interpret the national interest so as to increase viewership rather than express fidelity to history. Hence, news can turn into myth. A good example was the Anna Hazare campaign in 2011, which Times TV's coverage propelled to the number one news story, defining it very powerfully in for-or-against terms. Previously, the English-language media had always positioned popular and grassroots campaigns as threats to law and order. With Anna Hazare, that was neatly inverted. It was critics who were rendered suspect, while support became the norm. Support for what? That was less clear. Hazare's campaign was against corruption, defined in such broad terms as to become a nameless evil everyone could oppose. When it came time to negotiate political differences, the media euphoria was switched off, and the campaign rapidly turned into politics as usual. The campaign proved that even in a fragmented media environment, commercial television's propaganda power easily dwarfed that of the state when the latter monopolized broadcasting (see Rajagopal 2011).

It is not surprising, in this context, that the state appears more like a character actor on television than its guiding force, following the American model, although state actors evoke a good deal less reverence than in the US. The debates on ownership and control and programming policy that dominated scholarship on Indian television until the 1990s have left little trace in the current literature. The attitude of business, which was once alternately timid and defiant before the government, is now confident and impatient. The Ministry of Information and Broadcasting, once dictatorial and peremptory, and invoking fear from industry leaders, now pleads for guidance so that it can better help business. Oppositions that figured so prominently in earlier debates between the national and the global, between state and market, and between public and private, have all nearly dissolved when we enter the terrain of the media, or have taken very different forms that remain to be properly understood. It is against such amnesia that the chapters in this volume need to be read, calling attention to a history whose traces are actively being covered over at present with independent television programming.

The Present History of TV Studies

We cannot imagine India today without television, writes Nalin Mehta in this volume: it has become indispensable to any account of contemporary India.[9] Arguably, one could not have made the same claim for print, or radio, or even the cinema, without evoking some amount of doubt and disagreement. Audience size in each case would have constrained claims about influence for those earlier media. In the case of television, however, the spread has been so rapid and extensive, and it occupies so much time and space, that the claim demands to be entertained. Half the population or more watches television at home. If television is, in fact, indispensable to any account of India today, how does it shape that account? What is the story of television in India, beyond recounting its spread and dwelling on its content?

A popular response television's ubiquity provokes is to applaud it, and to observe that here is a clear case of capitalist technological growth that the people want, and prosper from. Why else would people go to the expense to obtain it when basic necessities can easily exhaust the incomes of most households? Amidst so many reasons for gloom and despair, the fact that people turn to television is surely cause for optimism. People learn new ways of being, conceive of connections between their lives and those of others, and project mental horizons that can ameliorate their lives. The chapters in this volume decline the temptation of such an uncritical response.

The chapters can broadly be divided into three groups for the purposes of my discussion. The first group, which most pointedly reflects on the meta-discursive conditions of Indian television studies as a field, expresses concerns subsumed in the first part of this chapter (namely, the essays by Asthana, Kumar, Mankekar, and Roy). The second group (chiefly Sinha, Thussu, Mehta, and Hutnyk) seeks to characterize the kind of politics taking shape with the expansion of the television industry and of viewership in India. They are attentive to the fact that the ground of reference for determining the political has shifted, and thus part of their challenge is to redraw the coordinates for assessing the political, in the context of globalization and the retreat of the state. The level at which questions of the political are assessed are national and global, in these chapters; to this extent, questions of region and culture are implicitly treated as the content, while politics is granted the status of form.

The third group (chiefly, Sen, Chakrabarti, Gupta, and Vangal) invert the tacit hierarchy in the formulation of the previous group. Focused at the level of region and culture, they show how the formal understandings at the level of the national and global are confounded or toppled when we take specificities of language, region, and culture into account. Rather than 'culture' being the content and 'politics' the form, they show that any assertion of hierarchy (form being determining of content, for example) enfolds the likelihood of its subversion. I will address the themes of these two groups next.

Locating the Political

Globalization has led to an alteration of the forms through which state power is exercised. The media are not the centre of a critical theory of the media, Oskar Negt has written (Negt 1978: 63). Television is not exceptional or exemplary (Biswarup Sen), but it is clearly an important avenue through which new modes of exercising power are being practised. Political parties distribute free television sets; businesses and political parties vie for control over television channels; and the manufacture of propaganda has evolved to accommodate its mediation by private business. Meanwhile, the success of Indian television signals not the positive strength of the state as before, but its restraint, in 'allowing' the economy to grow (Rajagopal 2012).

What emerges in the chapters in this volume is that, with the formal retreat of the state to aid the market and the popularity of commercial television, there is no obvious sense of how to order the foci of scholarly concern. Globalization rules, but it manifests through localization. Meanwhile, ruling elites increasingly assume that the most meaningful goal for the state is to promote private enterprise, regarding public interest as a synonym for corruption. Audiences seem to prefer reality TV, soap operas, and certain styles of newscasting, but the issue of how to read these preferences is still clearly the site of ongoing scholarly inquiry. Given Indian television's recency, it is understandable that even defining the spectrum of scholarly disagreement is difficult at this time.

As long as broadcasting remained a government monopoly, the view that shaped much of the scholarship was: let broadcasting be free! Defining freedom in negative terms was simpler than determining what its positive content should be. Few scholars departed from the government's own prescription for what the medium ought to do (education, information, and entertainment), and criticisms were usually attributed to 'misuse', as in the Janata Party government's 1977 white paper on the media during the Emergency (Government of India 1977).

The emphasis on utility is less noticeable in these chapters, although they retain a keen awareness of political economy and of the prevalence of arbitrary practices of control, especially at the regional level in India. An earlier model of reception studies tended to treat producers and viewers of television programmes as functionally interchangeable, and television as a transparent device in cultural reproduction (for example, Katz and Liebes 1990). Questions of technology, and related issues of affect, intention, and temporality, have shifted the terrain of theory from humanism, to ask how new modes of subjectivation can be linked to the complex techno-political fields being created.

The existence of a television audience suggests a hypothetical middle ground between exposure to welfare and warfare, one that the state had earlier occupied and has now largely disavowed. But when businesses that now occupy the commanding heights point to market demand as shaping their decisions, and the market is a field of contingent choice, where and how is the political to be located? The keenness in these chapters about the medium's political implications is counterbalanced by the absence of an obvious way to anchor the political, between state welfare directed at the poorest sections of the society and what can only be called warfare strategies, including slum demolition, land allocation, and counter-insurgency operations (for example, in adivasi-dominated areas).

In the era of Doordarshan, television can be read through James Scott's argument about 'seeing like a state', where censored and vetted news shows and entertainment were screened for audiences. To adopt a different view, television and audiences are different 'state effects', to invoke another political theorist, Philip Abrams. For Abrams (1988), it is misleading to treat the state as an actually existing entity; rather, the state has to be aggregated from the series of effects it produces. The series of state effects is itself an incoherent series, providing no logic or rationale, and confirming the existence of no raison d'etat. The latter view, in its refusal of a substantialist theory of the state, implies a radical critique of James Scott's account. Scott assumes precisely what Abrams rejects, namely, an irreducible character the state is unable to escape, and asserts that certain classes of phenomena are uniquely statist. With a mass medium like television, able to capture the attention of a majority at any given time, perhaps we can say that *television looks like a state*—in at least two senses. First, what is *viewed on television* constitutes authoritative knowledge of a kind that, until recently, could only be provided by the state. This authority is retained even in the absence of state monopoly, it can be argued. Second, *viewing television* continues to be akin to viewing an apparatus of the state, albeit via the intermediary supervision of corporations. Media houses remain largely obedient to state diktat, and while they maybe critical of political parties or leaders, they are clearly effective whenever popular opinion has to be mobilized. The media have also demonstrated that they help extend the market and thereby act as revenue farmers for the state, since when the consumption of packaged goods increases, tax collection also increases.[10]

The Problem of Culture

Even after the explosion of private television, Doordarshan, the national broadcasting agency, retains the largest terrestrial television network, although inspection of available audience metrics would not easily disclose this fact.[11] Nielsen, the global ratings firm, controls the surveys that determine television ratings in India, and allegedly under-reports Doordarshan in its surveys of media audiences. This may reflect a widespread perception amongst advertisers that the state-owned television, with its poorer and more rural viewers, is barely relevant to marketers (see Sinha in this volume).

Previously, national and regional identities were common sense referents within a closed domestic economy, which the government regulated under an import-substitution industrialization regime. Audiences today experience programmes from the US or the United Kingdom (UK), recast in domestic formats, as well as shows that acknowledge traditional and conservative Indian values albeit in upscale, consumerist settings (Kumar, Roy, and Sen in this volume). With the opening of the economy and the exponential growth in its technological mediation, both regional and national identity appear, ironically, as the aim and outcome of global capital and the talent it employs.

When economic reforms began to be carried out, the task of the private sector was to advance the growth of the economy. What was the task of the media industry once it was granted freedom by the state, besides growing and making money? As it turned out, the government wanted little more than revenue growth. Meanwhile, after decades of warning against corrosive Western values, foreign investment flooded into the Indian media industry. The result, however, was in some respects the opposite of what had been feared: Indian-language programming, often expressing conservative values and upholding traditional roles, became among the most widely watched programmes. As it turned out, cultural identity was not only the basis for expanding the media market but also, in a way, its result. Corporations sought to brand goods so as to encourage old and new consumers to imagine intimacy with these goods. Cultural identity defined in regional, religious, or national terms served this purpose.

Identities necessarily changed in the process, into something more public and stereotyped, but reflexive and individuated as well; that is, they became harder externally and more plastic internally. While a measure of turbulence resulted from these new expressions of identity, the political order revealed itself to be surprisingly resilient.

Television, for long a symbol of the advanced character of Western modernity, has simultaneously reproduced Western insularity, through broadcasting that cultivated audiences according to its prevailing norms of improvement or entertainment. Its programming culture paid little attention to societies elsewhere, relegating them to the periphery of popular imagination. The term 'emerging markets', applied to many of the world's oldest market societies, reflects this strange conceit.

But with the globalization of media, it is not the West so much as the rest of the world where scholars seek to discern the future to come. We might say the difference is that between being and becoming. The West seeks to preserve the broad character of its present structures, while the rest of the world's population is aware that their future lies in transcending their present, not in preserving it.

Notes

1. Some of what follows appears in revised form in *Public Culture*, vol. 25, no. 3, August 2013, under the title, 'Putting America in its Place'.

2. See Martin Heidegger (2008); Roland Barthes [1988 (1977]; Friedrich A. Kittler (1999); Jean Baudrillard (1995).

3. For further discussion, see Arvind Rajagopal (2006a). A different view from Deleuze's might link rather than separate geographical areas, so that there was a need to connect analyses of regimes of control across the world, just as their production regimes are in fact linked, such as FoxCon's factories in China making iPhones and iPads with apparently little concern for worker safety.

4. For a more extended discussion of this portion of my argument, see Rajagopal 2013.

5. In this discussion, I am indebted to Paul Starr (2004). For discussion, see Rajagopal (2006b).

6. Marx's discussion here pertains to civil society, but his point, I believe, has wider application.

7. A recent version of this mythic structure can be found at work in McKenzie Wark's (2004) book. In this book, the Internet, which was conceived as a security measure against nuclear war by the US Department of Defense, and is subject to censorship and surveillance on an Orwellian-scale today, becomes an arena where radical computer programmers can transform society together. One hopes they will, but to assign them something akin to the revolutionary potential that Marx assigned the proletariat, when the limits on their criticism do not inform them, is strange to say the least.

8. In a conversation reported by the novelist Namita Gokhale (see Auletta 2012: 54).

9. The only technology that has spread faster is the cell phone, which is a far larger industry in terms of revenue, with a broader base and lower unit cost of sale. See Robin Jeffrey and Assa Doron (2012).

10. The felicitous phrase 'looking like a state' is from Steven Pierce (2006).

11. TAM India, a joint venture of Nielsen and Kantar Media, is at present the only electronic rating agency functioning in India. The principal method for calculating target rating points or TRP (indicating viewership of a channel or programme) is frequency monitoring, in which 'people meters' are installed in sample homes and these electronic gadgets continuously record data about the channel watched by the family members. Currently, there are 8,160 people meters installed across 148 million television households, leading to charges of under-reporting. At the time of writing (October 2012), NDTV and Doordarshan have both sued Nielsen for alleged manipulation of viewership data. In this connection, see Sanjiv Kumar Baruah (2012). Doordarshan has its own diary-based ratings system, DART (Doordarshan Audience Ratings). Viewers are asked to note down each programme as and when watched by family members; diaries filled out in this way are collected and the results tabulated by Doordarshan.

References

Abrams, Philip. 1988. 'Notes on the Difficulty of Studying the State', *Journal of Historical Sociology*, 1(1): 58–91.

Anderson, Benedict. 1983. *Imagined Communities: Reflections on the Origin and Spread of Nationalism*. London and New York: Verso.

Auletta, Ken. 2012. 'Citizens Jain: Why India's Newspaper Industry is Thriving', *The New Yorker*, 8 October, pp. 52–61.

Barthes, Roland. 1988 (1977). 'The Death of the Author', in *Image, Music, Text*. tr. Stephen Heath. New York: Noonday Press.

Baruah, Sanjiv Kumar. 'Nielsen Research Firm for Planned TV TRP System', *Hindustan Times*, 14 August, available at http://www. hindustantimes.com/India-news/NewDelhi/Nielsen-research-firm-for-planned-TV-TRP-system/Article1-913677.aspx (last accessed on 12 October 2013).

Baudrillard, Jean. 1995. *The Gulf War Did Not Take Place*. tr. Paul Patton. Bloomington, IN: Indiana University Press.

Chakrabarty, Dipesh. 2000. *Provincializing Europe: Postcolonial Thought and Historical Difference*. Princeton, NJ: Princeton University Press.

Deleuze, Gilles. 1992. 'Postscript on Control Societies', October, 59(Winter): 3–7.

Government of India. 1977. *White Paper on Misuse of Mass Media During The Internal Emergency*. New Delhi: Government of India Press.

Heidegger, Martin 2008. 'Letter on Humanism', in *Martin Heidegger: Basic Writings*. Harper Perennial Modern Classics.

Jeffrey, Robin and Assa Doron. 2012. *The Great Indian Phone Book*. Cambridge, MA: Harvard University Press.

Katz, Elihu and Tamar Liebes. 1990. *The Export of Meaning: Cross-cultural Readings of Dallas*. New York and London: Oxford University Press.

Kittler, Friedrich A. 1999. *Gramophone, Film, Typewriter*. tr. Geoffrey Winthrop-Young and Michael Wutz. Stanford, CA: Stanford University Press.

Laslett, Peter. 1970, *Locke's Two Treatises of Government: A Critical Edition with Introduction and Notes*, 2nd edition. Cambridge: Cambridge University Press.

Marx, Karl. 1844. 'On the Jewish Question', *Deutsch-Französische Jahrbücher*, available at http://www.marxists.org/archive/marx/works/1844/jewish-question/ (accessed on 21 October 2012).

Masani, Mehra. 1976. *Broadcasting and the People*. New Delhi: National Book Trust.

McLuhan, Marshall. 2011 (1964). *Understanding Media: The Extensions of Man*. Routledge Classics.

Negt, Oskar. 1978. 'Mass Media: Tools of domination or instruments of liberation? Aspects of the Frankfurt Schools' Communications Analysis,' *New German Critique*. no. 14, Spring 1978, p. 63.

Pierce, Steven. 2006. 'Looking Like a State: Colonialism and the Discourse of Corruption in Northern Nigeria', *Comparative Studies in Society and History*, 48(4): 887–914.

Rajagopal, Arvind. 2006a. 'Imperceptible Perceptions in Our Technological Modernity', in Wendy Hui Kyong Chun and Thomas Keenan (eds), *New Media Old Media: A History and Theory Reader*, pp. 277–86. New York: Routledge.

———. 2006b. 'An American Theory of the Public Sphere', *Sociological Forum*, 21(1): 147–57.

———. 2011. 'Visibility as a Trap in the Anna Hazare Campaign', *Economic and Political Weekly*, 46(47): 19–21.

———. 'Wanted: A Communications Policy', *The Hindu*, 24 January, available at http://www.thehindu.com/opinion/lead/wanted-a-communications-policy/article2826250.ece (last accessed on 6 October 2013).

———. 2013. 'Putting America in Its Place', *Public Culture*. August v 25, n. 3, pp. 387–99.

Starr, Paul. 2004. *The Creation of the Media: Political Origins of Modern Communication*. New York: Basic Books.

Tocqueville, Alexis de (ed. and trans. by Harvey Mansfield and Delba Winthrop). 2011[1835, 1840]. *Democracy in America*. Chicago: University of Chicago Press.

Touraine, Alain (trans by David Macey). 1995. *Critique of Modernity*. Cambridge, MA, and Oxford, UK: Blackwell.

Wark, McKenzie. 2004. *A Hacker Manifesto*. Cambridge, MA: Harvard University Press.

Index

Editors and Contributors

Sanjay Asthana is Associate Professor in Journalism at the Middle Tennessee State University. He earned his PhD in journalism and mass communication in 2003 from the University of Minnesota. He also holds an MPhil in Philosophy and an MA Communication from the University of Hyderabad in India. His major research areas include globalization and media, visual communications, postcolonial theory, and cultural studies. He is the author of *Youth Media Imaginaries from Around the World* published by Peter Lang in 2012, and is currently completing an international research grant/project on Israeli-Palestinian working-class youth media practices. He has also co-authored, 'Media Information Literacy: Policy and Strategy Guidelines', a report published by UNESCO in 2013. He is available at http://www.mtsu.edu/mcgrad/faculty/asthana_rec.php.

Santanu Chakrabarti is currently Head of Insights at Oxfam, Great Britain, and a part of the strategic planning team for the organization. In his academic work he studies how culture articulates with wider socio-economic and political processes. Among his current objects of inquiry are Hindu nationalism and the political economy of audience measurement. His latest paper, appearing in the journal *Media, Culture & Society*, is entitled 'How Structure Shapes Content, or Why the "Hindi turn" of Star Plus Became the "Hindu turn"'

Nilanjana Gupta is Professor of English and former Director, School of Media, Communication and Culture, at Jadavpur University. She was also the Dean of the Arts Faculty at Jadavpur. Her current research project is titled 'Acropolis Now: India's New Urban Frontier' and focuses on the emergent structures of urban life in Rajarhat New Town, a 38 sq km development on the eastern border of Kolkata, and Gurgaon, a satellite town south of Delhi.

John Hutnyk is the author of *The Rumour of Calcutta* (Zed Books, 1996); *Critique of Exotica* (Pluto Press, 2000); *Bad Marxism: Capitalism and Cultural Studies* (Pluto Press, 2004); *Diaspora and Hybridity* (Sage Publications, co-authored with Raminder Kaur Kahlon and Virinder Kalra 2006), and *Pantomime Terror: Music and Politics* (Zero Books, 2014) plus several edited volumes including *Dis-Orienting Rhythms: The Politics of the New Asian Dance Music* (Zed Books, co-edited with Sanjay Sharma and Ashwani Sharma 1996) and *Beyond Borders* (Pavement Books, 2012). He blogs at Trinketization http://hutnyk.wordpress.com.

Shanti Kumar is an Associate Professor in the Department of Radio–Television–Film and a faculty affiliate in the Department of Asian Studies, the Center for Asian-American Studies, and the South Asia Institute at the University of Texas, Austin. Before joining the University of Texas in 2006, he taught at the University of Wisconsin–Madison and the University of North Texas. He is the author of *Gandhi Meets Primetime: Globalization and Nationalism in Indian Television* (University of Illinois Press, 2006) and the co-editor of *Planet TV: A Global Television Reader* (New York University Press, 2003). He has also written columns and short essays for online journals such as Flowtv.org, *Antenna*, and *In Media Res*.

Purnima Mankekar is an Associate Professor at the Departments of Gender Studies and Asian American Studies, University of California, Los Angeles. Her research is in feminist television and media studies, postcolonial theory, interdisciplinary theories of affect, and transnational cultural studies. Her book, *Unsettling India: Affect, Temporality, Transnational Public Cultures*, is forthcoming with Duke University Press. She has two books in preparation, one on counter-publics and 'September 11, 2001', and a second on television in post-liberalization India. She is currently conducting ethnographic fieldwork (in collaboration with Akhil Gupta) in call centres and the business process outsourcing (BPO) industry in Bangalore.

Nalin Mehta is Visiting Senior Fellow at Asia Research Institute and Institute of South Asian Studies, National University of Singapore, and joint editor of the international journal, *South Asian History and Culture* (Routledge), as well as the Routledge 'South Asian History and Culture' book series. He is an Adjunct Professor at the Indian Institute of Management, Bangalore, and has been a senior analyst with the United Nations and the Global Fund in Geneva, Switzerland, for the past four years. He started his career as a political journalist with NDTV and was deputy news editor and anchor at Times Now.

Arvind Rajagopal is a Professor in the Department of Media, Culture and Communication, and an affiliate faculty in the Department of Sociology, and the Department of Social and Cultural Analysis, at New York University. His book, *Politics after Television* (Cambridge University Press, 2001), won the Ananda Kentish Coomaraswamy Prize from the Association of Asian Studies in 2003. His recently published articles include essays in *Bioscope, Modern Asian Studies, Public Culture* and *SAMAJ*. In 2010–11, he was a Fellow at the Center for Advanced Study in the Behavioral Sciences at Stanford University. He is currently completing a book on post-independence political culture in India.

Abhijit Roy is Associate Professor of Film Studies at Jadavpur University, Calcutta. A member of the editorial board of *Journal of the Moving Image*, Roy has published widely on a number of articles on television, popular culture, and politics, and has written a book on Sergei Eisenstein (Papyrus, Calcutta, 2004). His recent articles include, 'A Reflexive Turn in Television Studies? Conjectures from South Asia' (*Journal of South Asian History and Culture*, 2012) and 'The Border Within: India and Manipur', in John Hutnyk (ed.), *Beyond Borders* (Pavement Books, London, 2012). He has received Visiting Fellowship from Maison des Sciences de l'Homme (Paris, 2006) and Visiting Scholarship from Indian Institute of Advanced Study (Shimla, 2009).

Biswarup Sen is Assistant Professor in the School of Journalism and Communication at the University of Oregon. His research interests include global media studies, reality television, new media, and communication theory. He has authored, *Of the People: Essays on Indian Popular Culture* (Chronicle Books, 2006), and has contributed chapters to *Global Bollywood: Transnational Travels of Hindi Music* (Minnesota, 2006) and *Global Formats: Understanding Television across Borders* (Routledge, 2012). He is currently working on a book-length project that examines the impact of the digital revolution on contemporary Indian culture.

Dipankar Sinha is Professor and Head, Department of Political Science, University of Calcutta. He is also Honorary Visiting Professor at the Institute of Development Studies, Kolkata, Honorary Associate of the Centre for Media History, Macquarie University, Sydney, and Nominated Member, Association of Third World Studies, USA. His area of interest broadly relates to the communicative modes of globalization, governance, and development. He specifically concentrates on exploring politics of communication vis-à-vis media and new technology.

Daya Kishan Thussu is Professor of International Communication and Co-Director of India Media Centre at the University of Westminster in London. He has a PhD in International Relations from Jawaharlal Nehru University, New Delhi. Founder and Managing Editor of the Sage journal, *Global Media and Communication*, his latest book is, *Communicating India's Soft Power: Buddha to Bollywood* (Palgrave/Macmillan, 2013).

Uma Vangal is Professor, Department of Direction and Head of the Media and Entertainment Department at LV Prasad Film & TV Academy, Chennai, and visiting faculty at the School of Media Studies, Loyola College. She is also an independent filmmaker, critic, and journalist whose work includes several short feature films and documentaries as well as ad films and promotional material for the corporate sector. She regularly conducts workshops and training sessions designed for media students, teachers, and professionals.